Lecture Notes in Networks and Systems

Volume 293

The series "Lecture Notes in Networks and Systems" publishes the latest developments in Networks and Systems—quickly, informally and with high quality. Original research reported in proceedings and post-proceedings represents the core of LNNS.

Volumes published in LNNS embrace all aspects and subfields of, as well as new challenges in, Networks and Systems.

The series contains proceedings and edited volumes in systems and networks, spanning the areas of Cyber-Physical Systems, Autonomous Systems, Sensor Networks, Control Systems, Energy Systems, Automotive Systems, Biological Systems, Vehicular Networking and Connected Vehicles, Aerospace Systems, Automation, Manufacturing, Smart Grids, Nonlinear Systems, Power Systems, Robotics, Social Systems, Economic Systems and other. Of particular value to both the contributors and the readership are the short publication timeframe and the world-wide distribution and exposure which enable both a wide and rapid dissemination of research output.

The series covers the theory, applications, and perspectives on the state of the art and future developments relevant to systems and networks, decision making, control, complex processes and related areas, as embedded in the fields of interdisciplinary and applied sciences, engineering, computer science, physics, economics, social, and life sciences, as well as the paradigms and methodologies behind them.

Indexed by SCOPUS, INSPEC, WTI Frankfurt eG, zbMATH, SCImago.

All books published in the series are submitted for consideration in Web of Science.

More information about this series at http://www.springer.com/series/15179

Dorota G. Pijanowska · Krzysztof Zieliński ·
Adam Liebert · Janusz Kacprzyk
Editors

Biocybernetics and Biomedical Engineering – Current Trends and Challenges

Proceedings of the 22nd Polish Conference on Biocybernetics and Biomedical Engineering, Warsaw, Poland, May 19–21, 2021

Springer

Editors
Dorota G. Pijanowska
Polish Academy of Sciences
Nalecz Institute of Biocybernetics
and Biomedical Engineering
Warsaw, Poland

Krzysztof Zieliński
Polish Academy of Sciences
Nalecz Institute of Biocybernetics
and Biomedical Engineering
Warsaw, Poland

Adam Liebert
Polish Academy of Sciences
Nalecz Institute of Biocybernetics
and Biomedical Engineering
Warsaw, Poland

Janusz Kacprzyk
Systems Research Institute
Polish Academy of Sciences
Warsaw, Poland

ISSN 2367-3370 ISSN 2367-3389 (electronic)
Lecture Notes in Networks and Systems
ISBN 978-3-030-83703-7 ISBN 978-3-030-83704-4 (eBook)
https://doi.org/10.1007/978-3-030-83704-4

This Springer imprint is published by the registered company Springer Nature Switzerland AG
The registered company address is: Gewerbestrasse 11, 6330 Cham, Switzerland

Contents

Automatic Assessment of Benton Visual Retention Test Results: A Pilot Study

Dominika Gabor[1], Rafał Doniec[2], Szymon Sieciński[2(✉)], Natalia Piaseczna[2], Konrad Duraj[2], and Ewaryst Tkacz[2]

[1] Evertop sp. z o.o., Działkowa 8, 41-506 Chorzów, Poland
dominika.gabor@onet.eu
[2] Department of Biosensors and Processing of Biomedical Signals, Silesian University, Faculty of Biomedical Engineering, Roosevelta 40, 41-800 Zabrze, Poland
{rafal.doniec,szymon.siecinski,natalia.piaseczna,konrad.duraj, etkacz}@polsl.pl

Abstract. Visual memory plays an important role in learning to read, write, spell, or draw. The dysfunction of visual memory can be detected by applying tests, such as Benton Visual Retention Test (Benton Test, BVRT), which is based on displaying ten patterns for ten seconds per each pattern and asking the subject to reproduce the displayed patterns. The purpose of our study was to automate the Benton Test by developing a desktop application. We automated the assessment of hand-drawn geometrical shapes (triangles, circles, rectangles) by applying machine learning and image processing techniques, namely, the ResNet50 network trained for recognition of triangles, circles, rectangles and other shapes, filling in the shape, and determining the type of triangle. The proposed algorithm proved its reliability on subjects with limited ability to draw shapes and was the part of a desktop application which may find its use in screening visual perception and visual memory.

Keywords: Benton Visual Retention Test · Image recognition · Automatic assessment

1 Introduction

The brain is a part of the central nervous system which controls the work of the human body. It controls basic life functions such as breathing or eating, body movements, communication, expressing emotions, and also observation and analysis of the environment, memorization and learning [8]. Visual memory plays an important role in remembering what we see [10].

The appropriate level of visual perception enables a person to learn to read, write, and spell [14]. We do not pay much attention to the visual memory until its

Electronic supplementary material The online version of this chapter (https://doi.org/10.1007/978-3-030-83704-4_1) contains supplementary material, which is available to authorized users.

D. G. Pijanowska et al. (eds.), *Biocybernetics and Biomedical Engineering – Current Trends and Challenges*, Lecture Notes in Networks and Systems 293,
https://doi.org/10.1007/978-3-030-83704-4_1

dysfunctions affect everyday activities [3,6]. There are many diagnostic methods for the assessment of visual memory impairment [1] but in our study, we used Benton Visual Retention Test.

Benton Visual Retention Test (Benton Test, BVRT) is a visual perception and visual memory test published in 1946 by Arthur Lester Benton [9] administered to children, adolescents, and adults [2,10,11]. During the test, the examined individual is shown ten patterns, one at a time, and then asked to reproduce each pattern exactly from memory. The time of the test is unlimited and the results are scored by form, shape, pattern, and arrangement. The invention of BVRT was based on the experiences of psychologist Arthur Lester Benton and neurologist Morris Bender in the treatment of servicemen suffering from traumatic brain injuries [5,9,13].

The aim of the study is to automate the evaluation of the results of the Benton Visual Retention Test by developing a desktop application.

2 Materials and Methods

2.1 Experiment Setup

The study group consisted of 9 residents at the nursing home in Borowa Wieś, a neighborhood of Mikołów (Poland), who can read, write, and draw. Based on the level of disability, they were assigned to three categories: people with diagnosed mild (A), moderate (B), and severe disability (C), as shown in table 1. The degree of disability of each subject was determined by primary care facilities upon admission to the nursing home.

Table 1 The groups of subjects

Group	Subjects
Mild disability (A)	2, 4, 8
Moderate disability (B)	1, 5, 7
Severe disability (C)	3, 6, 9

Each subject was asked to participate in Benton Test which consisted of the following steps:

1. Displaying ten patterns of basic geometrical shapes (triangles, rectangles, circles) or a blank space which appeared in corners;
2. Remembering the pattern in 10 s, regardless of the degree of intellectual disability;
3. Reproducing the pattern without time limit;
4. Repeating steps 1–3 ten times;
5. Evaluating the reproduced pattern.

The geometrical shapes in our implementation of Benton test included circles, triangles, rectangles, and blank shapes (no shape) presented in the four corners

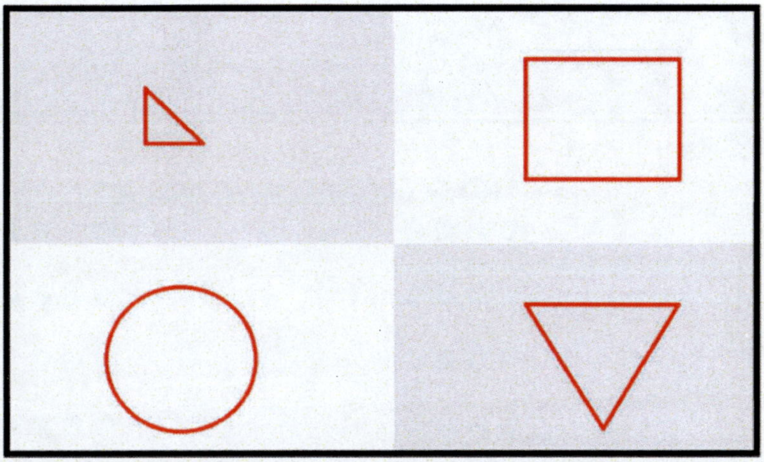

Fig. 1 An example of pattern to be reproduced by a subject

of the pattern. An example of a pattern to be reproduced during the BVRT is shown in Fig. 1.

The experiment outline was approved by the Bioethics Committee of the Medical University of Silesia in Katowice (The resolution KNW/0022/KB1/18) and carried out following the rules of the Declaration of Helsinki of 1975, revised 2013.

2.2 Assessment of Results

The ability to reproduce the pattern as accurately as possible is measured by the number of errors found in the reproduced shapes. These errors are divided into six categories (omission, distortions, repetition, shifts, rotations, and wrong size) indicating disorders of visual perception [2].

In our study, we automate the assessment of geometrical shapes reproduced by a subject by preprocessing and applying deep neural network (DNN) to recognize them. Then, the errors are classified based on the following principles:

- The omission occurs when a subject did not draw a geometrical shape in a designated area.
- Wrong size means that the reproduced geometrical shape is significantly smaller or larger than in the presented pattern.
- Distortion occurs when the reproduced figure does not reflect the sample (e.g., a triangle instead of a rectangle).
- Rotation error is defined reproducing a shape in a direction which does not reflect the sample.

The next step is listing the detected errors and displaying them to the user and the set of actual and correct answers with the occurrence of a type of error (Fig. 2).

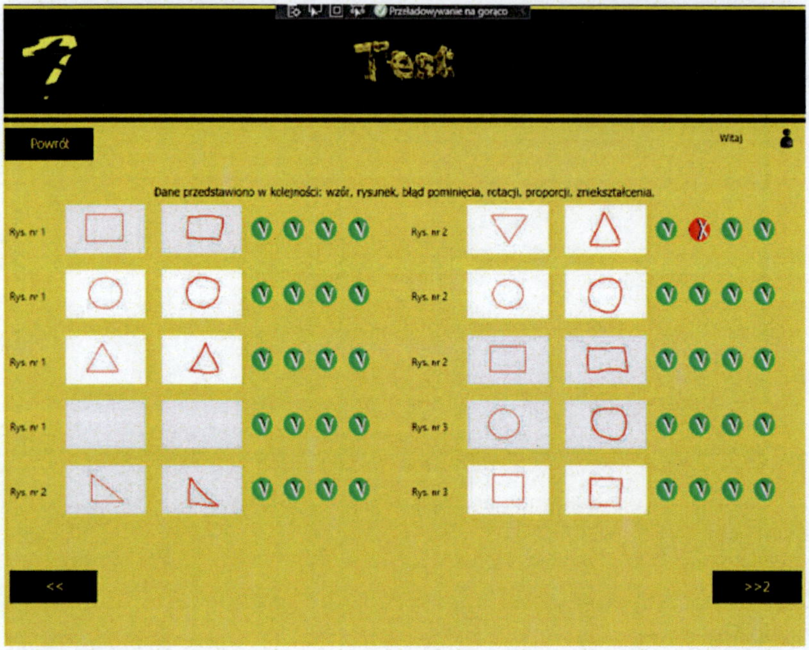

Fig. 2 The indication of errors in a graphical manner. The tick (✓) mark indicates the correct reproduction whereas the cross (✗) mark indicates the occurrence of error (omission, rotation, size, distortion)

2.3 Image Processing

Recognition of shapes consists of preprocessing and applying the DNN trained on 40 samples of triangles, rectangles, circles, and other figures painted in red. Each geometrical shape was represented by 10 samples. The preprocessing consists of changing the color of the background to white and filling in the area with red color because the lines are red, as in the training set. After this operation, the geometric shape area was calculated.

The geometrical shapes after preprocessing are fed to a classifier based on deep artificial neural network (DNN). The DNN was implemented in C# which uses TensorFlow to implement transfer learning [7] using ResNet50 network described in detail in [4]. The output of the DNN was the label describing a shape ("p" for the rectangle, "t" for a triangle, "o" for circle, and "i" for other shape). Then, the triangles were divided into six categories.

Triangle Type Coding. Because the DNN could not recognize the rotation of a triangle, we proposed a new approach to address this problem; the triangle was divided into nine parts by three horizontal and three vertical lines (see Fig. 3). If most pixels in the horizontal axis (from right to left) were in the first part, the figure was given the number 100 (1 which represents the first part with the maximum number of pixels and zero (0) for other parts). A similar rule applies for the vertical axis (from top to bottom). The codes and triangle rotations are presented in Fig. 4.

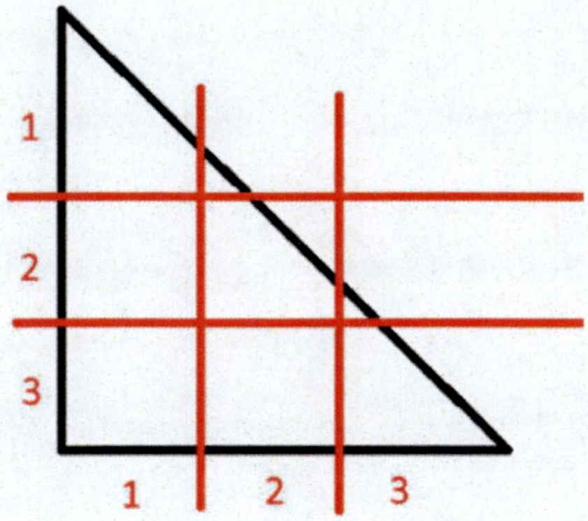

Fig. 3 The division of a triangle into nine parts

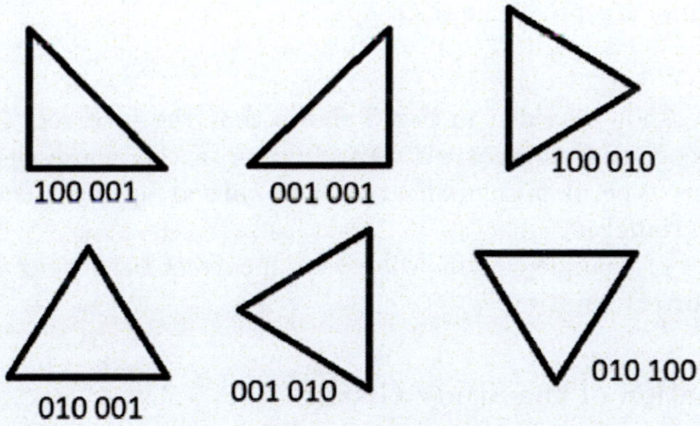

Fig. 4 Triangle rotation codes

3 Results

3.1 Shape Recognition

The results of BVRT were evaluated based on the description published in [2]. The number of correctly recognized rotations, distortions, changes of size, omissions, triangles, circles, and rectangles was 10 out of 10 evaluated geometrical shapes for each category. The number of correctly recognized incorrect rotations was 6. Figure 5 shows an example of misclassification of rotation errors.

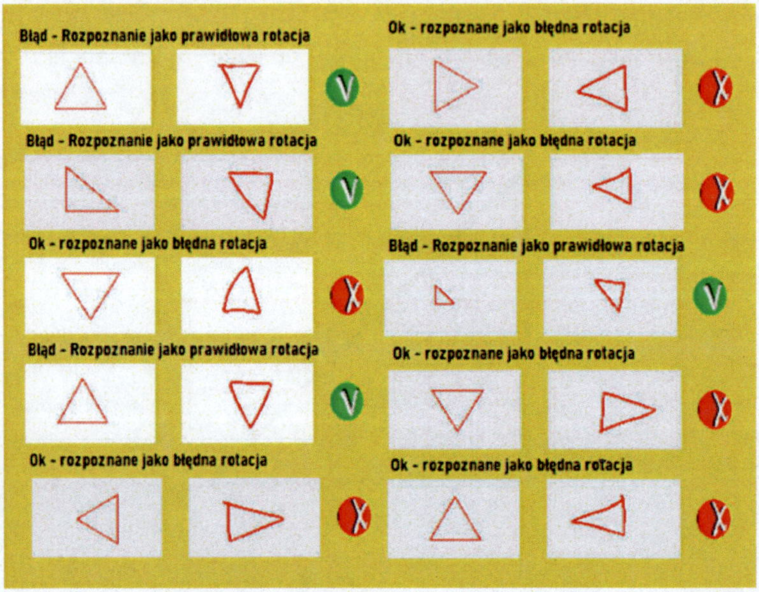

Fig. 5 Examples of correctly and incorrectly recognized geometrical shapes and the anomalies (errors in reproducing the shape)

The screenshot provided in Fig. 5 shows that the incorrect rotation in four cases was recognized as a correct reproduction of the shape (left image in the column). This type of misclassification was caused by incorrect coding of the hand-drawn triangles.

Four cases of triangle classification were incorrect because of incorrect recognition of a correct rotation.

3.2 Evaluation of the Study Group

The Table 2 shows the number of errors (divided by their category) committed in each group category. It can be observed that in the group of mildly disabled people, which was significantly different from the group of moderately and severely disabled people, there is also a difference in the type of errors made. In group A, the most common were rotation errors, and in groups B and C – distortions and omissions.

4 Discussion

The purpose of our study was to automate the evaluation of Benton Visual Retention Test by developing an algorithm and a desktop application. We proposed an algorithm based on image processing and deep neural network implemented in C# using ML.NET and Tensorflow libraries and ResNet50 neural

Table 2 Number of errors made in each group

Error category	Group A	Group B	Group C
Distortion	3	13	15
Omission	0	13	12
Rotation	8	2	5
Shifts	4	2	1
Wrong size	0	5	2

network. The proposed algorithm recognizes the geometrical shapes drawn by a subject and then classifies them as a triangle, circle, or rectangle.

The next step is verifying whether the shape drawn by the examined person is consistent with the pattern. The proposed algorithm proved its reliability on subjects with limited ability to draw shapes in a careless and sloppy manner. However, in some cases, the algorithm classified the geometric shapes incorrectly because some hand-drawn geometric shapes such as a circle, square, or triangle may look quite similar [12].

Another disadvantage of the proposed algorithm is the incorrect recognition of the triangle rotation in some cases, which works properly only for basic rotations, presented in Fig. 4. Nevertheless, the proposed algorithm can recognize the hand-drawn geometrical shapes (triangles, rectangles and circles) and errors in reproducing them which may be used to assess the dysfunctions of memory and recommend the appointment with a psychologist.

Since the intelligence index affects visual perception, and this is closely related to the degree of impairment, we expected to notice a large difference between the three study groups. Nevertheless, a significant difference in the test results was visible between the group of mildly disabled people and the other two groups. On the other hand, the outcomes in groups B and C were similar. The result obtained in this way suggests that visual perception in people with different degrees of impairment is not different and there are no significant differences between them.

The most significant limitations were the small number of subjects in the study group (only 9 people) and incorrect recognition of triangle rotations in some cases. In future work, we consider including the recognition of more geometrical shapes, their modifications (distortions, rotations, and scales), and extending the study groups by including more subjects.

References

1. Baldwin, S., Farias, S.T.: Neuropsychological assessment in the diagnosis of Alzheimer's disease. Current Protoc. Neurosci. **49**(1) (2009). https://doi.org/10.1002/0471142301.ns1003s49
2. Benton, A.L.: A visual retention test for clinical use. Arch. Neurol. Psychiatr. **54**(3), 212–216 (1945). https://doi.org/10.1001/archneurpsyc.1945.02300090051008

3. Gersztenkorn, D., Lee, A.G.: Palinopsia revamped: a systematic review of the literature. Surv. Ophthalmol. **60**(1), 1–35 (2015). https://doi.org/10.1016/j.survophthal.2014.06.003
4. He, K., Zhang, X., Ren, S., Sun, J.: Deep residual learning for image recognition. In: 2016 IEEE Conference on Computer Vision and Pattern Recognition (CVPR), pp. 770–778, June 2016
5. Kreutzer, J., Caplan, B., DeLuca, J.: Encyclopedia of Clinical Neuropsychology, vol. 4. Springer, New York (2010). https://doi.org/10.1007/978-0-387-79948-3
6. McTighe, S.M., Cowell, R.A., Winters, B.D., Bussey, T.J., Saksida, L.M.: Paradoxical false memory for objects after brain damage. Science **330**(6009), 1408–1410 (2010). https://doi.org/10.1126/science.1194780
7. Microsoft Corporation: VisionCatalog. ImageClassification Method. Microsoft Docs (2021). https://docs.microsoft.com/en-us/dotnet/api/microsoft.ml.visioncatalog.imageclassification?view=ml-dotnet#Microsoft_ML_VisionCatalog_ImageClassification_Microsoft_ML_MulticlassClassificationCatalog_MulticlassClassificationTrainers_Microsoft_ML_Vision_ImageClassificationTrainer_Options_. Accessed 28 Jan 2021
8. NICHD - Eunice Kennedy Shriver National Institute of Child Health and Human Development: What are the parts of the nervous system? October 2018. https://www.nichd.nih.gov/health/topics/neuro/conditioninfo/parts. Accessed 27 Jan 2021
9. Sivan, A.B., Levin, H.S., Hannay, J., Benton, A.L.: Pioneer, colleague, mentor, and friend. J. Int. Neuropsychol. Soci. **13**(04) (2007). https://doi.org/10.1017/s1355617707070774
10. Study.com: Visual memory: Definition & skills, December 2016. https://study.com/academy/lesson/visual-memory-definition-skills.html. Accessed 25 Jan 2021
11. Talarowska, M., Florkowski, A., Zboralski, K., Gałecki, P.: Wykonanie Testu Pamięci Wzrokowej Bentona oraz Wzrokowo-Motorycznego Testu Gestalt Lauretty Bender przez osoby z depresją i organicznymi zaburzeniami depresyjnymi [Results of the Benton Visual Retention Test and the Bender Visual-Motor Gestalt Test among patients suffer from depressive disorders and organic depressive disorders]. Psychiatria Polska **45**(4), 495–504 (2011). http://cejsh.icm.edu.pl/cejsh/element/bwmeta1.element.desklight-15462a2c-4dce-49a4-8b3b-813b69f7e67c/c/ZN_Ped_2015_11_Szeligiewicz_Urban_Danuta.pdf
12. Verdine, B.N., Lucca, K.R., Golinkoff, R.M., Hirsh-Pasek, K., Newcombe, N.S.: The shape of things: the origin of young children's knowledge of the names and properties of geometric forms. J. Cogn. Dev. **17**(1), 142–161 (2015). https://doi.org/10.1080/15248372.2015.1016610
13. Walsh, W., Betz, N.: Tests and Assessment. Prentice Hall (1990). https://books.google.pl/books?id=FhK0HvTN8PoC
14. Łysoń, A.: Ćwiczenia percepcji i pamięci wzrokowej, September 2011. http://www.publikacje.edu.pl/pdf/8977.pdf. Accessed 25 Jan 2021

Analysis and Comparison of Heart Rate Variability Signals Derived from PPG and ECG Sensors

Aleksandra Królak(✉) and Edyta Pilecka

Institute of Electronics, Lodz University of Technology, Łódź, Poland
aleksandra.krolak@p.lodz.pl

Abstract. One of the methods of diagnosis and prevention of cardiovascular diseases is monitoring and analysis of heart rate variability (HRV). However, in order to accurately detect the RR intervals, the ECG signal must be acquired. The advances in optical sensors made photoplethysmography (PPG) to be considered as an alternative for heartbeat interval measurements as it is more convenient. The aim of this study is to compare the HRV signals calculated from ECG and PPG measurements based on the data from two Physionet datasets and assess the reliability and accuracy of time-domain, frequency-domain and nonlinear HRV parameters derived from the ECG and PPG signals. The results achieved suggest that the PRV can be used as an alternative for HRV analysis in static measurements, with significant correlation coefficient values above 98% for nonlinear, time and frequency features. In the dynamic measurements, performed during sports activities, especially nonlinear and frequency-domain parameters derived from PPG signals should be used with caution, as in some cases the correlation was inverse.

Keywords: ECG · PPG · HRV · PRV · HRV parameters

1 Introduction

Despite significant progress in medicine, mortality from cardiovascular diseases is very high. As stated by Eurostat, in 2017 accounted for 41.4% of deaths in Poland, which places these diseases in the first place in terms of the cause of death. Globally, this value in 2016 reached the level of 31%, which was the most common cause of death in the world for a given year. These data confirm that cardiovascular diseases are a serious problem on a global scale, and that the diagnosis and prevention of these ailments are still a field for analysis and improvement. One of the methods of diagnosis and prevention of cardiovascular diseases is monitoring and analysis of cardiac activity. Heart Rate Variability (HRV) refers to the heart rate of the heart muscle and describes the difference in intervals between the R points of consecutive QRS complexes. This is a result of the autonomic nervous system functions and its influence on the work of the heart muscle. HRV shows how the cardiac muscle adjusts its work to external stimuli. Increased heart rate variability indicates an increased activity of the parasympathetic

D. G. Pijanowska et al. (eds.), *Biocybernetics and Biomedical Engineering – Current Trends and Challenges*, Lecture Notes in Networks and Systems 293,
https://doi.org/10.1007/978-3-030-83704-4_2

nervous system, and decreased activity of the sympathetic nervous system. In case of measurement of pulse-to-pulse distances from photoplethysmography (PPG) the signal is sometimes related as pulse rate variability (PRV). The great advantage of PRV is the fact that it can be measured using low-cost, off-the-shelf devices such as fitness bands or smartphones [1, 2].

HRV analysis can be a useful tool to detect cardiac diseases [3], as well as in diagnosis of depression [4] or in measuring sports performance [5]. HRV parameters, such as low-frequency (LF) and high frequency (HF) modulation of R–R interval changes, standard deviation of all normal–normal (N–N) intervals (SDNN), the root mean square of standard deviations between adjacent N–N intervals (RMSSD) and Standard Descriptors 1 and 2 (SD1 and SD2) [6], are more and more often used to set optimal training loads during training [7, 8].

Because the mechanisms and physiological aspects of ECG and PPG vary, a number of studies have been conducted to see if PPG is adequate as an HRV assessment technique. According to previous research, most of the presented results suggests that estimation of HRV from PPG is sufficient [9] or even that there exists high correlation between HRV and PRV [10, 11]. Significant correlation between peak-to-peak intervals was also reported in [12]. The results presented in [13] showed that the PRV spectrum almost perfectly matches HRV, particularly in healthy subjects at rest. Physical activity, on the other hand, may reduce the correlation, particularly for the high frequency (HF) component. The findings in [14] suggest that using PPG to assess HRV properties should be carefully considered, particularly for wearable devices and health care information systems. Jeyhani et al. showed, that comparing HRV calculated from PPG and ECG from 19 healthy subjects, the relative error was equal to ~30% [15]. Pinheiro et al. suggests that PRV can be used as an alternative for HRV analysis in healthy individuals, however time and frequency domain parameters should be used with caution in post-exercise and CVD subjects [16].

The aim of this study is to compare the HRV signals calculated from ECG and PPG measurements based on the data from two Physionet datasets and assess the reliability and accuracy of time-domain, frequency-domain and nonlinear HRV parameters derived from the ECG and PPG signals.

2 Materials and Methods

2.1 Datasets

For analysis of HRV and PRV two datasets from Physionet were used [17]: Wrist PPG During Exercise Database [18] and BIDMC PPG and Respiration Dataset [19].

Wrist PPG dataset contains the ECG and PPG waveforms obtained as a result of measurements made during the performance of individual forms of physical exercise for a specified period of time:

- Walking on the treadmill for up to 10 min
- Jogging or easy running on the treadmill for up to 10 min
- Pedaling a stationary bike with a low load for up to 10 min
- Pedaling a stationary bike with a high load for up to 10 min

The group of respondents consisted of 8 people—3 men and 5 women, aged 22 to 32 years—the average age was 26.5 years. The subjects could choose the pace and duration of the test so as not to exceed their comfort limit. In most cases, the exercise time was 4 to 6 min. The ECG signal was recorded using the CamNtech Actiwave device. The PPG signal was recorded with a measuring unit which, in addition to the Shimmer 3 GSR optical sensor, consisted of a high-accuracy accelerometer, gyroscope and magnetometer. The measuring apparatus was placed on the left wrist of the subject. The sampling frequency for all recorded measurements was 256 Hz. The PPG signal was not filtered for gait or smooth running measurements, while for measurements performed on an exercise bike, the signals were filtered with a Butterworth low-pass filter with a cut-off frequency of 15 Hz. Regarding the ECG signal, all measurements were filtered with a 50 Hz notch filter. In this work the first 3 min of the recorded signals were analyzed.

The data stored in BIDMC database was acquired from critically-ill patients during hospital care at the Beth Israel Deaconess Medical Centre (Boston, MA, USA). The recordings within the dataset, each of 8-min duration, contain: PPG, impedance respiratory signal and ECG, sampled at 125 Hz, as well as heart rate (HR), respiratory rate (RR), and blood oxygen saturation level (SpO_2), sampled at 1 Hz. In this work the first 3 min of 10 selected signals from BIDMC dataset were analyzed.

2.2 Data Analysis

The analysis of the data from Physionet datasets was performed according to the algorithm implemented in Python and presented in Fig. 1. R-peak detection in ECG signal and pulse detection PPG signal was done using standard peak detection method with adjusted parameters. HRV signal parameters were calculated using hrv0.2.8. Python module [20]. HRV parameters selected for statistical analysis were: SDNN (standard deviation of N–N intervals), RMSSD (root mean square differences of successive R–R intervals), SDSD (standard deviation of successive differences between adjacent NN intervals), ratio of SD1 and SD2 (standard deviations of Poincare scattergram), Total Power (over all frequency bands), LF (low-frequency modulation of R–R interval changes) and HF (high frequency modulation of R–R interval changes). The statistical analysis was done using Analysis Toolpack in Microsoft Excel.

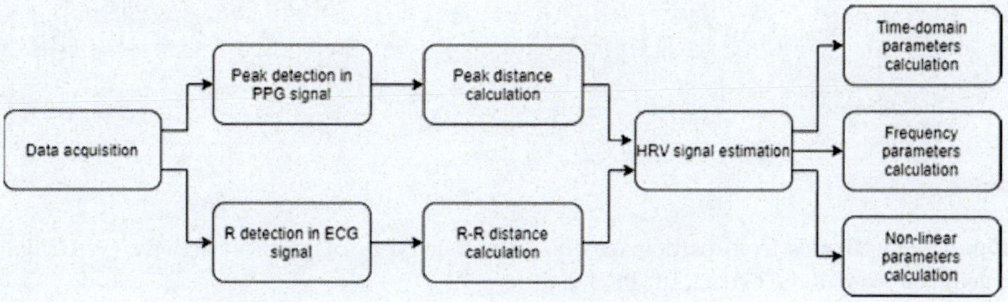

Fig. 1 Data processing procedure for HRV parameters estimation

3 Results

The synchronization of PPG and ECG data with detected R-peaks and pulse peaks are presented in Fig. 2a for Wrist database and in Fig. 2b for BIDMC database. It can observed that signals from Wrist database are not synchronized, while in case of BIDMC the detected heart beats are overlayed.

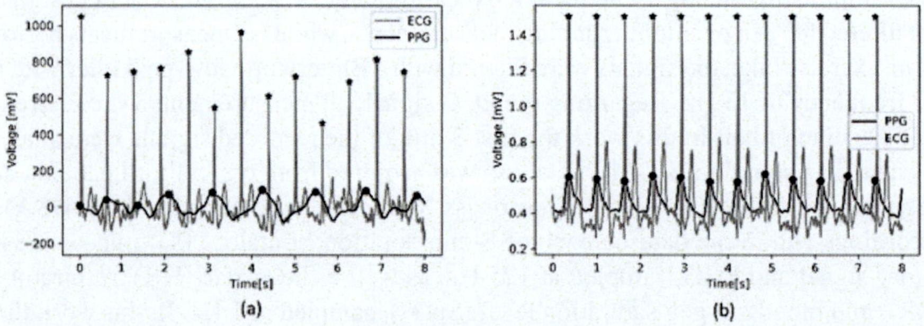

Fig. 2 Synchronization of ECG and PPG data: (a) for Wrist dataset, (b) for BIDMC dataset

The calculated HRV and PRV signals for data from Wrist database for the only participant who performed all 4 forms of activity are presented in Fig. 3a and Fig. 3b. It can be observed that in case of HRV signal (Fig. 3a) the levels of the signals suggest that the most demanding task was pedaling on an exercise bike set at a higher resistance, next was running on a treadmill, pedaling on an exercise bike set at a low resistance, and the least demanding activity was walking. For the calculated PRV signals for the same participant (Fig. 3b) it is visible that the order of the signals' levels is different: R-R intervals values show that pedaling on an exercise bike set at a low resistance was most demanding, while running appeared to be the least demanding activity.

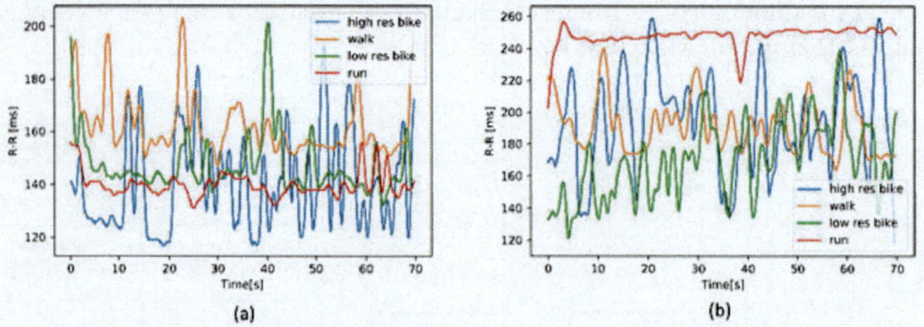

Fig. 3 HRV signals from participant no. 3 for four types of physical activity (Wrist dataset) calculated from: a) ECG data, (b) PPG data

The HRV and PRV signals derived from BIDMC database are almost overlaying, as it can be seen in Fig. 4.

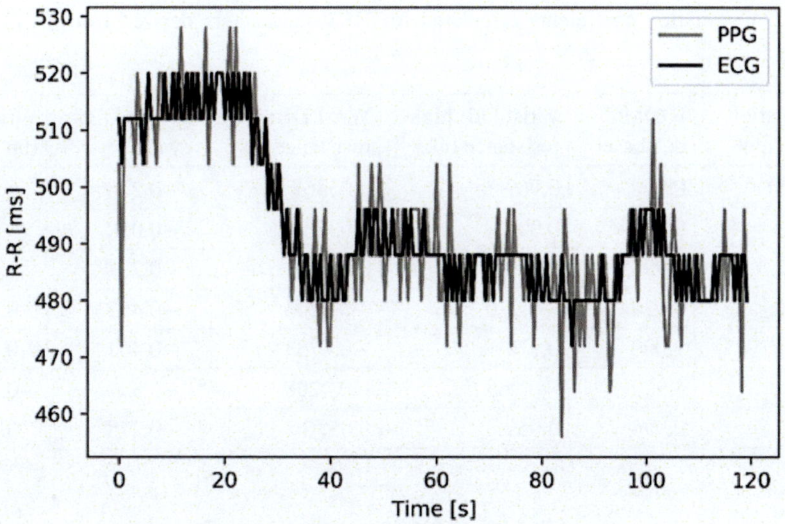

Fig. 4 HRV signals calculated from ECG and PPG data

Tables 1, 2 and 3 show calculated statistical parameters comparing the time-domain, frequency-domain and nonlinear HRV parameters for all types of analyzed signals. Table 1 presents values of p-value, in Table 2 the correlation coefficients for HRV and PRV parameters are given, and Table 3 shows the mean error for the HRV and PRV parameters calculated according to Eq. (1), where HRV_{param} denotes parameter value calculated from ECG signal and PRV_{param} is the parameter calculated from PPG signal.

$$Error = \frac{\left|HRV_{param} - PRV_{param}\right|}{HRV_{param}} \cdot 100\% \tag{1}$$

Table 1 p-value calculated for HRV parameters derived from ECG and PPG signals

p-value	BIDMC database	Wrist DB: high resistance bike	Wrist DB: low resistance bike	Wrist DB: walk	Wrist DB: run
RMSSD	0,85	0,39	0,04	0,24	0,74
SDNN	0,85	0,22	0,05	0,16	0,92
SDSD	0,85	0,39	0,04	0,24	0,74
S1/S2	0,79	0,10	0,33	0,66	0,72
TP	0,64	0,55	0,87	0,85	0,83
LF	0,74	0,23	0,90	0,94	0,58
HF	0,62	0,32	0,77	0,93	0,98

Table 2 Correlation coeffcients calculated for HRV parameters derived from ECG and PPG signals

Correlation coefficient	BIDMC database	Wrist DB: high resistance bike	Wrist DB: low resistance bike	Wrist DB: walk	Wrist DB: run
RMSSD	0,996	0,969	0,0536	−0,278	0,494
SDNN	0,992	0,996	0,0376	−0,030	0,302
SDSD	0,996	0,969	0,0546	−0,277	0,494
S1/S2	0,929	−0,970	−0,7203	−0,452	0,241
TP	1,000	0,970	−0,4753	−0,402	0,081
LF	0,999	−0,965	−0,4208	−0,397	−0,239
HF	0,996	0,969	0,0536	−0,278	0,494

Table 3 Mean error calculated for HRV parameters derived from ECG and PPG signals

Mean error	BIDMC database	Wrist DB: high resistance bike	Wrist DB: low resistance bike	Wrist DB: walk	Wrist DB: run
RMSSD	18,35%	27,82%	46,18%	62,08%	38,77%
SDNN	13,81%	49,86%	77,59%	67,96%	49,46%
SDSD	18,35%	27,90%	46,25%	62,10%	38,81%
S1/S2	13,61%	25,35%	32,84%	56,77%	15,32%
TP	65,52%	181,81%	252,80%	967,92%	154,31%
LF	60,40%	398,32%	1119,78%	1535,16%	129,41%
HF	128,65%	114,64%	207,33%	897,51%	278,43%

4 Discussion

Presented statistical results show high p-value for all HRV parameters calculated for critically-ill patients (BIDMC dataset), where the data was acquired from persons staying in horizontal position, with almost no movement. In case of Wrist dataset the lowest p-values were obtained for exercises performed on the exercise bike. It suggests that measurements of heart rate in dynamic conditions may have negative influence on the precision of the recorded PPG signal. The correlation coefficient is very close to 1 for all the parameters calculated for data from BIDMC database. For dynamic measurements of PPG and ECG (Wrist database) we can observe high correlation coefficient only for time-domain parameters calculated for exercises with pedaling a stationary bike with a high load. In case of nonlinear and frequency-domain parameters the correlation in most cases is inverse. This may be the result of insufficient synchronization of the ECG and PPG signals recorded during dynamic exercises, inaccurate optical pulse sensor or influence of movement on the accuracy of the PPG measurement.

The sampling rate of ECG and PPG signals has significant influence on the HRV metrics accuracy. In [21] it was reported that 250-Hz sampling frequency of ECG signal is acceptable for HRV analysis including non-linear and frequency domain metrics. For time-domain analysis 100-Hz sampling frequency would also be acceptable. These results imply that the results of frequency domain analysis performed in the presented study may have limited confidence, especially for BIDMC data. On the other hand, findings presented in [22] suggest that the sampling rate of PPG signal above 20 Hz is sufficient for analysis of PRV metrics.

As reported in the review presented in [23], the ECG and PPG methods yielded discrepancies of less than 6% for most HRV measures and 29.9% for pNN50. In the presented study the errors of analyzed HRV/PRV metrics are significantly larger, especially for frequency-domain parameters, exceeding several hundred percent. That may be the result of errors in peak detection for HRV signal determination. These results show that the values of the HRV and PRV parameters should not be compared directly. What is recommended is comparison of the trends of these parameters, especially in case of measurements during physical activity.

5 Conclusions

This study focuses on the evaluation of time-domain, frequency-domain and nonlinear HRV parameters calculated from ECG and PPG signals. In many previous studies that compared the use of PPG and ECG signals to assess HRV it was reported that PRV is sufficient to replace ECG for HRV assessment. The results presented in this work suggest that the pulse signal can be used to assess the PRV signal in subjects at rest. In case of measurements during performing physical exercises the results from wearable optical devices should be used with caution.

References

1. Peng, R.-C., Zhou, X.-L., Lin, W.-H., Zhang, Y.-T.: Extraction of heart rate variability from smartphone photoplethysmograms. Comput. Math. Methods Med. **2015**, 1–11 (2015)
2. Cho, Y., Julier, S.J., Bianchi-Berthouze, N.: Instant stress: detection of perceived mental stress through smartphone photoplethysmography and thermal imaging. JMIR Ment. Health **6**(4), e10140 (2019)
3. Hillebrand, S., Gast, K.B., de Mutsert, R., Swenne, C.A., Jukema, J.W., Middeldorp, S., Rosendaal, F.R., Dekkers, O.M.: Heart rate variability and first cardiovascular event in populations without known cardiovascular disease: meta-analysis and dose–response meta-regression. EP Europace **15**, 742–749 (2013)
4. Hughes, J.W., Stoney, C.M.: Depressed mood is related to high-frequency heart rate variability during stressors. Psychosom. Med. **62**, 796–803 (2000)
5. Makivić, B., Nikić Djordjević, M., Willis, M.S.: Heart Rate Variability (HRV) as a tool for diagnostic and monitoring performance in sport and physical activities. J. Exerc. Physiol. Online **16**, 103–131 (2013)
6. Dong, J.G.: The role of heart rate variability in sports physiology (review). Exp. Ther. Med. **11**, 1531–1536 (2016)

7. Oliveira, R., Leicht, A., Bishop, D., Barbero-Álvarez, J., Nakamura, F.: Seasonal changes in physical performance and heart rate variability in high level futsal players. Int. J. Sports Med. **34**, 424–430 (2013)
8. Plews, D., Laursen, P., Kilding, A., Buchheit, M.: Heart rate variability in elite triathletes, is variation in variability the key to effective training? A case comparison. Eur. J. Appl. Physiol. **12**, 3729–3741 (2012)
9. Schafer, A., Vagedes, J.: How accurate is pulse rate variability as an estimate of heart rate variability? A review on studies comparing photoplethysmographic technology with an electrocardiogram. Int. J. Cardiol. **166**(1), 15–29 (2013)
10. Leikan, G.M., Rossi, E., Cuadra Sanz, M., Delisle Rodríguez, D., Mántaras, M.C., Nicolet, J., Zapata, D., Lapyckyj, I., Nicola Siri, L., Perrone, M.S.: Evaluation of agreement between temporal series obtained from electrocardiogram and pulse wave. J. Phys. Conf. Ser. **705**(1), 012038 (2016)
11. Peng, R.C., Zhou, X.L., Lin, W.H., Zhang, Y.T.: Extraction of heart rate variability from smartphone photoplethysmograms. Comput. Math. Methods Med. **2015**, 1–11 (2015)
12. Lee, J., Kim, J., Shin, M.: Correlation analysis between electrocardiography (ECG) and photoplethysmogram (PPG) data for driver's drowsiness detection using noise replacement method. Procedia Comput. Sci. **116**, 421–426 (2017)
13. Lin, W.H., Wu, D., Li, C., Zhang, H., Zhang, Y.T.: Comparison of heart rate variability from PPG with that from ECG. In: The International Conference on Health Informatics. IFMBE Proceedings, vol. 42. Springer, Cham (2014)
14. Sahroni, A., Hassya, I.A., Rifaldi, R., Ul Jannah, N., Irawan, A.F., Rahayu, A.W.: HRV assessment using finger-tip photoplethysmography (pulserate) as compared to ECG on healthy subjects during different postures and fixed breathing pattern. Procedia Comput. Sci. **161**, 535–543 (2019)
15. Jeyhani, V., Mahdiani, S., Peltokangas, M., Vehkaoja, A.: Comparison of HRV parameters derived from photoplethysmography and electrocardiography signals. In: 2015 37th Annual International Conference of the IEEE Engineering in Medicine and Biology Society (EMBC), pp. 5952–5955 (2015)
16. Pinheiro, N., Couceiro, R., Henriques, J., Muehlsteff, J., Quintal, I., Goncalves, L., Carvalho, P.: Can PPG be used for HRV analysis? In: 2016 38th Annual International Conference of the IEEE Engineering in Medicine and Biology Society (EMBC), pp. 2945–2949 (2016)
17. Goldberger, A., Amaral, L., Glass, L., Hausdorff, J., Ivanov, P.C., Mark, R., Stanley, H.E.: PhysioBank, PhysioToolkit, and PhysioNet: components of a new research resource for complex physiologic signals. Circulation **101**(23), e215–e220 (2000)
18. Jarchi, D., Casson, A.J.: Description of a database containing wrist PPG signals recorded during physical exercise with both accelerometer and gyroscope measures of motion. Data **2**(1), 1 (2017)
19. Pimentel, M.A.F., Johnson, A.E.W., Charlton, P.H., Birrenkott, D., Watkinson, P.J., Tarassenko, L., Clifton, D.A.: Towards a robust estimation of respiratory rate from pulse oximeters. IEEE Trans. Biomed. Eng. **64**(8), 1914–1923 (2016)
20. Hrv's documentation. https://hrv.readthedocs.io
21. Kwon, O., Jeong, J., Kim, H.B., Kwon, I.H., Park, S.Y., Kim, J.E., Choi, Y.: Electrocardiogram sampling frequency range acceptable for heart rate variability analysis. Healthc. Inform. Res. **24**(3), 198–206 (2018)
22. Choi, A., Shin, H.: Photoplethysmography sampling frequency: pilot assessment of how low can we go to analyze pulse rate variability with reliability? Physiol. Meas. **38**, 586–600 (2017)
23. Shaffer, F., Ginsberg, J.P.: An overview of heart rate variability metrics and norms. Front. Public Health **5**(258), 1–17 (2017)

Check for updates

Analysis of the Mechanical Properties of Impact Absorbing Structures Used in Military Helmets

Kamila Wiśniewska[1], Aleksandra Jędrzejewska[1,2], Monika Ratajczak[1(✉)], and Tomasz Klekiel[1]

[1] Department of Biomedical Engineering, Faculty of Mechanical Engineering, University of Zielona Góra, 65-516 Zielona Góra, Poland
m.ratajczak@iimb.uz.zgora.pl
[2] Science and Technology Park of University of Zielona Góra, 66-002 Zielona Góra, Poland

Abstract. This article deals with the issues of head protection against dynamic effects. The current design of the helmets does not fully protect against the destruction of the soft tissues of the head resulting, for example, as a result of an impact. Damage to these tissues is usually caused by dynamic loads, and soldiers are one of the most vulnerable groups to craniocerebral injuries. Hence, the study undertook the development of a model and numerical research on determining the optimal design solution for materials absorbing impact energy.

The work analyzes proprietary solutions of tubular and honeycomb structures for four different construction materials. The conducted research allowed to determine the mechanical properties of the proposed geometries of the helmet's dynamic load absorbing layer and the possibility of their use for head protection. Various solutions were analyzed in terms of geometry and material characteristics. In order to obtain the desired features of the analyzed structures, optimal values of parameters such as diameter or wall thickness were searched for. Two structures were analyzed in terms of the use of different materials.

The performed numerical analysis showed that one of these structures exhibits special energy-absorbing properties, thanks to which it can be an effective solution for protecting the skull and soft tissues of the head against the effects of dynamic loads and impacts.

Keywords: Numerical simulation · Strength analysis · FEM · Head injuries · Energy absorbers

1 Introduction

Traumatic brain injury (TBI) is a major cause of death and disability in people around the world [1]. These injuries arise as a result of excessive mechanical overload, for example, traffic accidents, falls from a height, sports activities (cycling,

D. G. Pijanowska et al. (eds.), *Biocybernetics and Biomedical Engineering – Current Trends and Challenges*, Lecture Notes in Networks and Systems 293,
https://doi.org/10.1007/978-3-030-83704-4_3

roller-skating, American football) or combat operations [2,3]. However, it should be emphasized that in combat conditions, the incidence of TBI is much higher than in other cases [4]. Research by Clausen et al. [5] showed that 10.7% of the 3,877 surveyed soldiers reported having experienced at least one head injury during service. Based on modern technologies, it is possible to create more and more effective structures to protect the head. This is an important issue due to the fact that head injuries can significantly reduce the ability to perform current activities, work, sports, etc., which significantly worsens the quality of life. The accident causes a craniocerebral trauma. The probability of closed injuries is much higher than open injuries [1]. Head injuries can be divided into brain and skull injuries. The most common head injuries include craniofacial or cerebral fractures and soft tissue injuries. On the other hand, the most common brain injuries include, among others, hematomas, contusions and concussions. Brain injuries can lead to loss of consciousness and disturbance of various functions of the brain [1].

When the head in the helmet collides with a stationary obstacle at high speed, there are significant accelerations due to deceleration, as well as a change in the direction of the velocity vector due to rebound. The helmet protects the skull from breaking, but the accelerations and the forces and moments of inertia generated in this way cause the displacement of the brain tissue and its movements in the cerebrospinal fluid (CSF). Then the critical acceleration values are exceeded, the brain may hit the skull walls, and thus –its permanent damage [6]. Both the analyzed clinical cases and computer simulations indicate that the solutions currently used in the construction of helmets do not fully absorb the impact energy. The search for a solution that can better protect brain tissues is a motivation to try to improve such systems.

The helmet belongs to the personal protective equipment [1]. It should protect the forehead, back, sides, temples and top of the head. In addition, the helmet must fulfill the following functions:

1. In the event of a collision, the task of the helmet shell is to distribute the forces generated during the impact over the largest possible surface in order to reduce the possibility of fracture of the skull bones.
2. Absorbing impact energy, reducing acceleration affecting the head and brain, minimizing health complications for the user.
3. Maintaining an unchanged, stable position during standard use, especially during an accident, which is ensured by an appropriate retention system (fastening straps and fasteners).
4. Protection of the greater part of the head surface, protecting it against foreign objects, while ensuring optimal visibility and audibility [1].

The helmet should also reduce head-to-neck load transfer in the event of an accident to prevent whiplash injuries [1]. The helmet consists of four basic components: the outer shell (shell), the energy-absorbing lining, the comfort lining, and the fastening straps and fasteners [1,8] (Fig. 1).

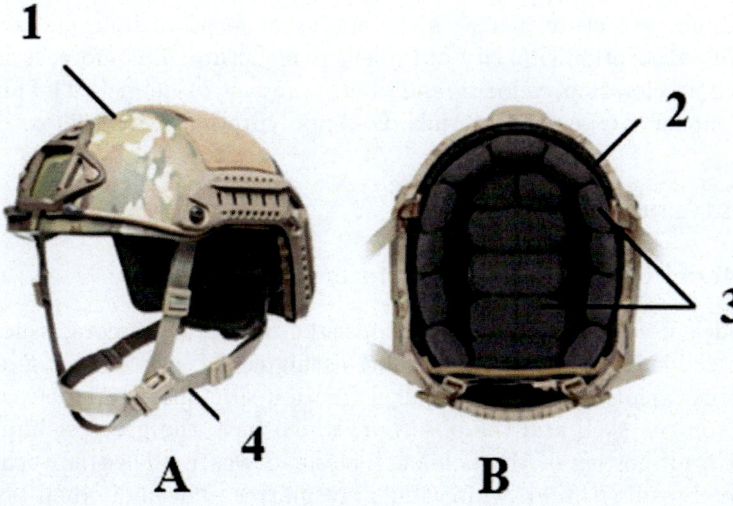

Fig. 1 Schematic construction of the helmet: **a** side view of the helmet; **b** view of the inside of the helmet; 1 - shell, 2 - protective lining, 3 - comfort lining, 4 - fastening straps and clasp; own elaboration based on [1,9]

Currently, one of the best methods of studying the biomechanical response of human tissues to mechanical loads are numerical methods such as the finite element method (FEM) [10,11]. Analyzes using this method also enable the search for appropriate design solutions that dampen the impact energy and protect the human body [12]. Hence, in this study, research was undertaken to improve the energy-consuming properties of helmet linings using FEM numerical tests.

The aim of the study was to determine the influence of selected shapes of absorbers on the protection of brain tissues. Two shapes, "tubular" and "honeycomb", were selected for the study, in which the destruction of the cellular structure and the friction between the walls of these structures were used as the effect of energy absorption. The "tubular" shape of the absorber is used, inter alia, in aviation, astronautics, the transport industry and for the construction of tanks for transporting liquids [13]. This structure has good energy absorbing properties and at the same time has a low weight due to the air content between the tubes. The source of inspiration for the shape of the "honeycomb" was the structure of bone tissue, but the proposed structure is more structured and has larger voids than bone tissue. This structure provides absorption properties due to the elastic properties. Pakzad et al. [14] have confirmed that the "honeycomb" structure increases the energy absorption time and can increase the energy absorption capacity of the entire structure. Therefore, it can be used to reduce acceleration, velocity and energy impact of helmet. The analysis was carried out for 4 types of materials: EVA, polyurethane, cork, flex.

2 Materials and Methods

2.1 Model Construction and Materials

The models used for the simulation consisted of 9 layers, among which the head and protective helmet elements can be distinguished, as well as a fixed layer on which other elements hit as a result of motion simulation. The elements of the helmet are: the shell and the absorber, which have the greatest impact on the protective properties of the helmet. The head was modeled as such layers as: skull, cerebrospinal fluid, white and gray matter. The individual layers of the model with the parameters are presented in Table 1. Several different materials were assigned for the absorber layer, as the study investigated the impact of both the absorber geometry and its construction material on stresses and deformations of the head tissues. The materials for the absorber layer were selected on the basis of Ashby diagrams [15].

The model layers were stacked on top of each other while remaining in contact without interfering. In addition to the fixed layer constituting the place of impact, the model layers were bent to reflect the natural curvature of the head and the helmet. Two models differing in the geometry of the helmet's absorber layer were tested. The first model is shown in Fig. 2, in this model the absorber layer has a tubular geometry. The second model with an absorber with a honeycomb geometry is shown in Fig. 3. The models were developed as head and helmet fragments, reducing the computation time.

2.2 Simulation Conditions

The value of stress, energy and strain values during a dynamic impact against a rigid, stationary, flat obstacle were adopted as a criterior for assessing the degree of impact of the load on the structures of the brain. The impact simulation was performed assuming the initial speed of 6 m/s given to the elements of the helmet and the head towards the stationary obstacle. The speed was selected on the basis of a literature analysis of FEM studies of protective helmets used by soldiers, motorcyclists and footbolists [16,17]. The simulation also took into account the effect of the force of gravity, which acted against the direction of speed.

Table 1 Material parameters of individual layers of models

Material	Model element	Type of material	Parameters*	Thickness [mm]
Steel 40H	Source of force	Linear isotropic	ρ = 7820 [9] E = 207 GPa [9] p_r = 0.3 [9]	10
Kevlar	Shell	Linear isotropic	ρ = 1440 [18] E = 58 GPa [18] p_r = 0.3 [18]	10 [19]
EVA	Absorber	Linear isotropic	ρ = 100 E = 20 MPa p_r = 0.45	30 [1]
Polyurethane	Absorber	Linear isotropic	ρ = 1000 E = 60 MPa p_r = 0.3	30 [1]
Cork	Absorber	Linear isotropic	ρ = 120 E = 26 MPa p_r = 0.26	30 [1]
Flex	Absorber	Linear isotropic	ρ = 1220 E = 72.9 MPa p_r = 0.31	30 [1]
Compact bone	Skull	Linear isotropic	ρ = 2000 E = 15000 MPa p_r = 0.22	7 [7]
Dura mater	Dura mater	Linear isotropic	ρ = 1140 [1] E = 31,5 MPa [1] p_r = 0.45 [1]	1.1 [20]
Pia mater	Pia mater	Linear isotropic	ρ = 1130 [21] E = 15 MPa [21] p_r = 0.45 [21]	7 [22]
Cerebro spinal fluid	Cerebro spinal fluid	Linear isotropic	ρ = 1000 [1] E = 0.15 MPa [1] p_r = 0.49999 [1]	2 [23]
Gray matter	Gray matter	Viscoelastic	ρ = 1060 [1] E = 4 kPa [1] p_r = 0.45 [1] Momentary shear modulus: G0 = 680 Pa [25] Viscoelastic constant: 80 s^{-1} [25]	2, 5 [24]
White matter	White matter	Viscoelastic	ρ = 1060 [1] E = 4 kPa [1] p_r = 0.45 [1] Momentary shear modulus: G0 = 1040 Pa [25] Viscoelastic constant: 80 s^{-1} [25]	25

ρ - density [kg/m^3], E - Young's modulus, p_r - Poisson's ratio

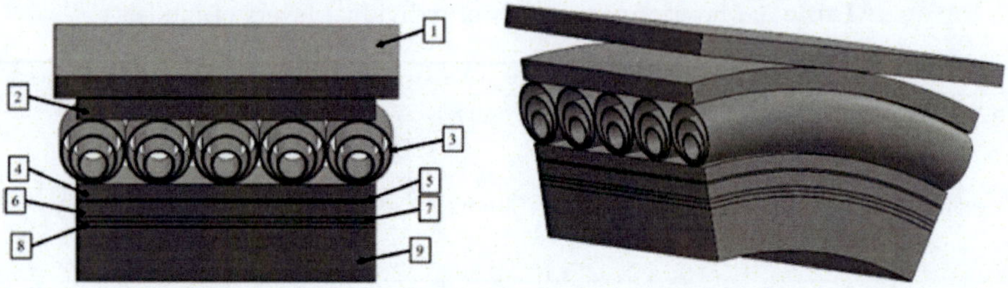

Fig. 2 Model for simulating the dynamic loads of the head with a tubular helmet absorber layer composed of the following layers: 1-force source, 2-shell, 3-absorber, 4-skull, 5-dura mater, 6-pia mater, 7-cerebro-spinal fluid, 8-gray matter and 9-white matter

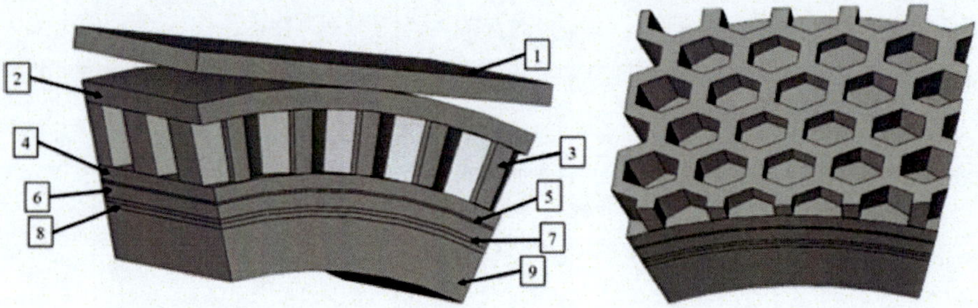

Fig. 3 Model for simulating the dynamic loads of the head with an absorber layer of a honeycomb helmet composed of the following layers: 1-force source, 2-shell, 3-absorber, 4-skull, 5-dura mater, 6-pia mater, 7-cerebro-spinal fluid, 8-gray matter and 9-white matter

3 Results

The stress distribution for the shapes of tubular and honeycomb absorbers as well as cork and flex materials is shown in Fig. 4. The highest stresses occurred for both shapes and all tested materials on the helmet shell. On the absorber layer, the absorption of energy formed on the helmet shell took place, thanks to which the tissue layers experienced less stress than on the shell, thus the proposed structures have an absorption potential. Below the absorber layer, the highest stresses occurred in the skull. In the case of the honeycomb absorber (Fig. 4A and B), a lower degree of energy absorption was observed for the flex material compared to the cork. This change is visible in Fig. 4B by increasing stresses on the lower layers of the model.

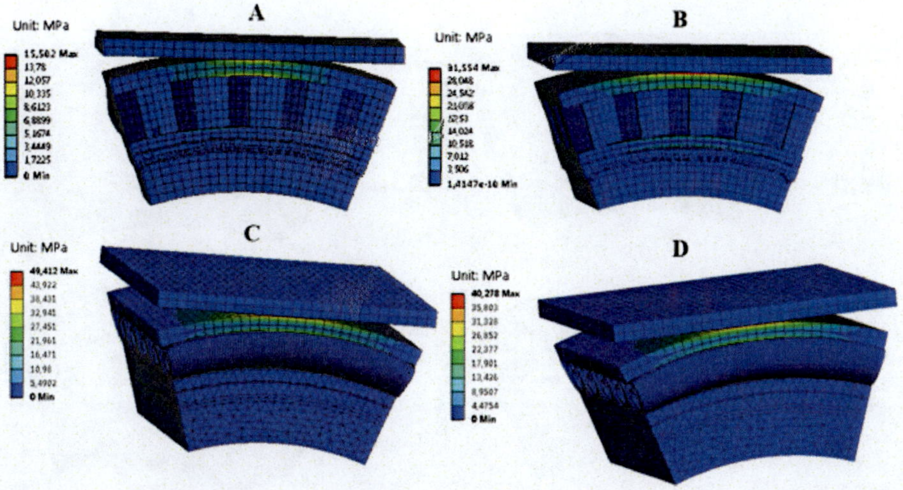

Fig. 4 Stress distribution for: the shape of the absorber A, B-honeycomb and C, D-tubular, materials A, C-cork, B, D-flex

Figure 5 shows the distribution of strains for the tubular absorber and honeycomb shapes made of cork and flex materials. For the honeycomb absorber, there were greater deformations of the tissues when using the flex material compared to the cork material (Fig. 5A and B), which is consistent with the stress distribution in Fig. 4. In the case of the tubular shape, the greatest deformations occurred for the absorber itself and increased after changing the material from cork to flex. The tubular absorber deforms more than the honeycomb absorber.

The results of numerical analyzes for stresses and strain of the skull (Fig. 6) and the white and gray matter of the brain (Figs. 7 and 8) caused by the impact allowed to determine the protective properties of the proposed structures of the absorber layers of helmets protecting the head. The phenomena occurring on the skull and the brain were analyzed, due to the fact that these structures are the most vulnerable to injuries, the effects of which may significantly reduce the quality of life of people who suffer an accident.

The presented graphs show the values of stresses, strain and their energies for the skull and brain as a result of an impact for various geometries and materials of structures that can be used for lining helmets. In the case of the honeycomb geometry, a large dependence of the strain value on the skull on the material of the absorber layer was observed, as shown in Fig. 6. In the case of tubular geometry, slight changes in the strain on the skull were observed when the absorber material changed. In the case of stresses and strains for the brain presented on Fig. 7, a smaller influence of the absorber material on the stress and strain values of the brain was observed. Studies have shown the occurrence of lower stresses and strains for both the skull and the brain when using the tubular shape of the absorber compared to the shape of the honeycomb. The tubular shape of the absorber therefore has better absorption properties.

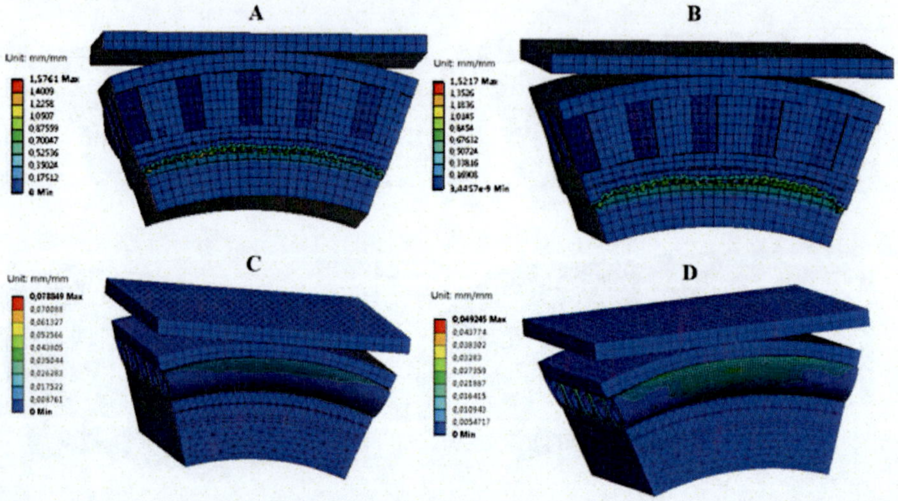

Fig. 5 Distribution of strain for: the shape of the absorber A, B-honeycomb and C, D-tubular, materials A, C-cork, B, D-flex

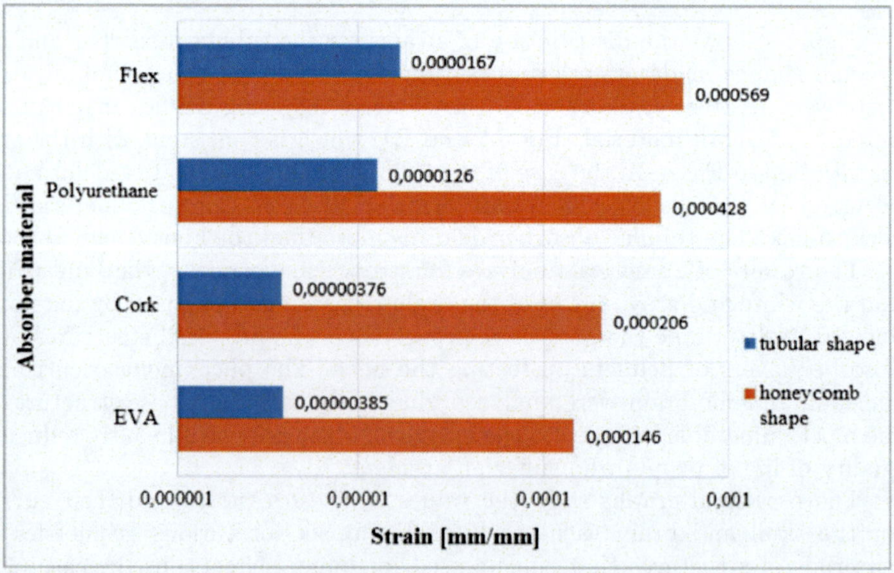

Fig. 6 Strain energy values on the skull resulting from the simulated impact depending on the geometry and material of the absorber

The low values of the strain of the brain correspond to the low values of the strain energy presented in Fig. 8 for both the examined structures and the absorber materials used.

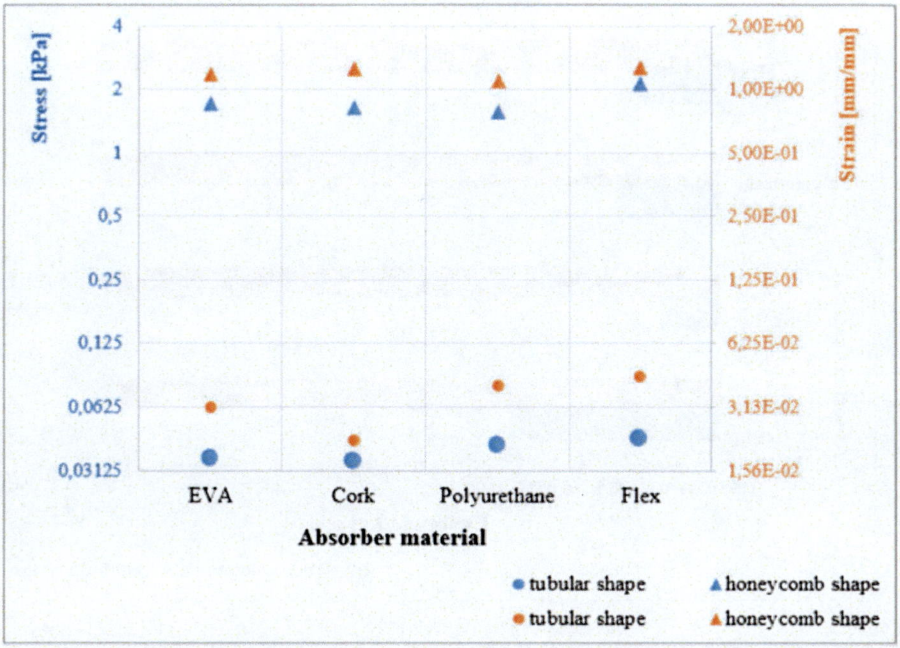

Fig. 7 Stress and strain values on the white and gray matter of the brain resulting from the impact simulation depending on the geometry and material of the absorber

4 Analysis of the Results

Based on the literature reports [25, 26], it was determined that the maximum stress on brain tissues that did not cause their damage was 34 kPa, and the strain was 0.2 mm/mm, while for the skull these values were respectively 346 MPa and 0.055 mm/mm. The honeycomb geometry does not provide sufficient protection of the skull against its damage, due to the excessive stiffness resulting from the structure. The protective properties of the honeycomb structure largely depend on its material, the best protection is provided by the EVA structure, worse properties are given by natural cork, while significant deterioration of properties was observed for materials such as polyurethane and flex. Based on the analysis of stress and strain on the skull, it was found that the best protection was provided by the tubular geometry, because the stress and deformation caused by the impact in the skull do not exceed the permissible values (346 MPa and 0.055 mm/mm).

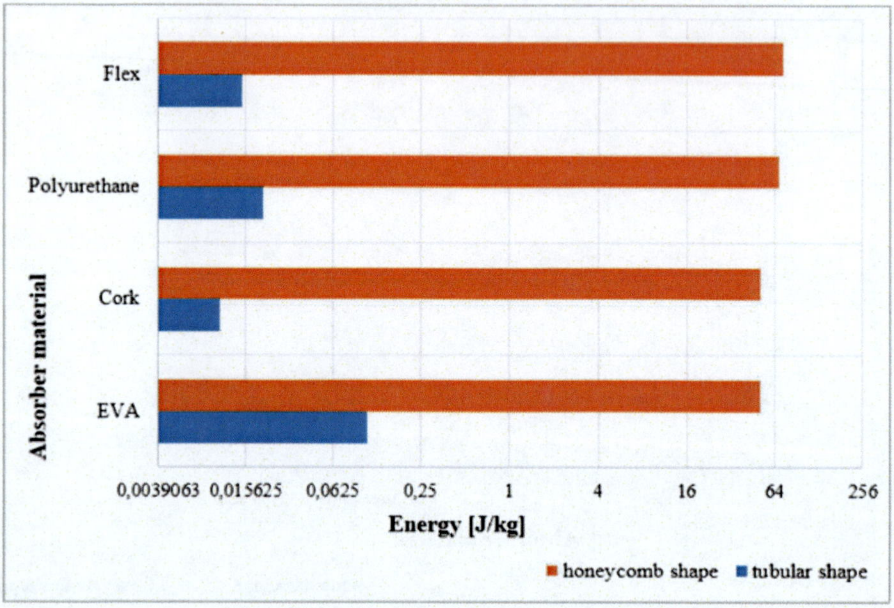

Fig. 8 Strain energy values on the white and gray matter of the brain resulting from the impact simulation depending on the geometry and material of the absorber

Both examined structures, regardless of their construction material, provide adequate protection of the white and gray matter of the brain, the stresses induced by the impact on these brain structures are lower than the permissible stress of 34 kPa. Similar conclusions were obtained by Galvanetto et al. [24] for EVA material, which confirms the rightness of using this material for the construction of the absorber and good absorption properties due to stresses on brain tissues. However, the honeycomb geometry does not provide sufficient protection against strain of the brain, as they amount to more than 0.2 mm/mm for these structures, i.e. they exceed the allowable strain.

5 Conclusions

Taking into account all the configurations, the best energy attenuation is provided by the absorber with a tubular structure, while the materials used do not include the best material for its construction. All tested materials can be used to construct a tubular absorber.

In the case of the honeycomb absorber, an influence of the type of material on its protective properties was observed. Depending on the construction material, this structure has different mechanical properties that affect the impact energy absorption capacity. The highest impact energy dispersion was observed for the honeycomb structure made of EVA material. Both proposed structures show good mechanical properties allowing the absorption of the impact energy, however, in the case of the honeycomb geometry, the stiffness is higher and this is insufficient to protect the skull from damage. This geometry would need to be structurally modified to improve its ability to absorb energy. The tubular geometry, on the other hand, has shown good protective properties and is potentially suitable for use as a helmet cushion layer due to the possibility of significant deformation during loading and thus converting the impact energy into work and friction work. The preliminary analysis of the structure of the helmet's damping elements presented in the paper will constitute the basis for further research on the search for the optimal structure, which requires the development of, among others, complete headform with helmet. Zhang and Sharma [27] for the model reconstructed from the real helmet obtained deformations lower than acceptable, the absorber used had a dense foam structure and was constructed of polyurethane, which suggests that the proposed structure will be more effective. The results of the research confirm that polyurethane is a material suitable for the construction of the absorber, and as shown in this study, the structure of the absorber has a significant impact on the ability to dissipate energy and thus prevent brain damage. In order to use the honeycomb structure, this structure should be appropriately modified in order to give it better absorption properties. The necessity to modify the structure in order to obtain optimal protection of the head is also confirmed by the works of Caserta et al. [28].

References

1. Radzicki, L.: Kaski rowerowe. Konstrukcja, technologie, użytkowanie. Wydawnictwo Politechniki Świętokrzyskiej, Kielce, pp. 49–89 (2015)
2. Ratajczak, M., Ptak, M., Chybowski, L., Gawdzińska, N., Będzinski, R.: Material and structural modeling aspects of brain tissue deformation under dynamic loads. Materials **12**(2), 1–14 (2019)
3. Dymek, M., Ptak, M., Ratajczak, M., Fernandes, F., Kwaitkowski, A., Wilheml, J.: Analysis of HIC and hydrostatic pressure in the human head during NOCSAE tests of American football helmets. Brain Sci. **11**(3), 1–29 (2021)
4. Ratajczak, M., Klekiel, T., Sławiński, G., Będziński, R.: Investigation of helmet-head interaction in the aspect of craniocerebral tissue protection. Adv. Intell. Syst. Comput. **1033**, 308–315 (2020)
5. Clausen, A.N., Clarke, E., Phillips, R., Haswell, C.: Combat exposure, posttraumatic stress disorder, and head injuries differentially relate to alterations in cortical thickness in military Veterans. Neuropsychopharmacology **45**, 491–498 (2020)
6. https://arxiv.org/ftp/arxiv/papers/2003/2003.07698.pdf (07.03.2021)
7. National Center for Biotechnology Information. http://www.ncbi.nlm.nih.gov
8. https://www.msf-usa.org/downloads/helmet_csi.pdf (07.03.2021)
9. https://www.intechopen.com/books/textiles-for-advanced-applications/contemporary-personal-ballistic-protection-pbp- (13.03.2021)
10. Klekiel, T., Arkusz, K., Sławiński, G., Będziński, R.: Prediction of the segmental pelvic ring fractures under impact loadings during car crash. Adv. Intell. Syst. Comput. **831**, 138–149 (2019)
11. Arkusz, K., Klekiel, T.: The influence of osteoporotic bone structures of the pelvic-hip complex on stress distribution under impact load. Acta of Bioeng. Biomech. **20**(1), 29–38 (2018)

12. **Arkusz, K.**, **Klekiel, T.**, Sławiński, G., Będziński, R.: Influence of energy absorbers on Malgaigne fracture mechanism in lumbar-pelvic system under vertical impact load. Comput. Methods Biomech. Biomed. Eng. **67**, 1–11 (2019)
13. Olabi, A.G., Morri, E., Hashmi, M.S.J., Gilchrist, M.D.: Optimised design of nested circular tube energy absorbers under lateral impact loading. Int. J. Mech. Sci. **50**, 104–116 (2008)
14. Pakzad, S., Ahad, S.H., Maryam, D., Amirbahadur, K.: Design and analysis of bicycle helmet made of PETg and ABS with honeycomb structure against external impacts using Abaqus software. Int. J. Adv. Des. Manuf. Technol. **14**, 83–90 (2021)
15. Ashby, M.F.: Materials Selection in Mechanical Design. Elsevier Ltd. New York, p. 251. ISBN 978-0-75-06616-3 (2005)
16. De-Shin, L., Yao-Te, Ch.: A finite element investigation into the impact performance of an open-face motorcycle helmet with ventilation slots. Appl. Sci. **7**, 279 (2017)
17. Sandberg, M., Tse, K., Long, B., Lee, H.P.: A computational study of the EN 1078 impact test for bicycle helmets using a realistic subject-specific finite element head model. Comput. Methods Biomech. Biomed. Eng. **21**, 1–9 (2018). https://doi.org/10.1080/10255842.2018.1511775
18. Cui, L., Forero, R.M.A., Gilchrist, M.D.: Optimisation of energy absorbing liner for equestrian helmets. Part I Layered Foam Liner Mater. Des. **30**, 3405–3413 (2009)
19. National Research Council. Review of Department of Defense Test Protocols for Combat Helmets. Washington, DC: The National Academies Press (2014). https://doi.org/10.17226/18621
20. Howay, L.: Characterization and Modification of Helmet Padding System to Improve Shockwave Dissipation. Clemson University, TigerPrints (2010)
21. Depreitere, B.: A rational approach to pedal cyclist head protection, PhD dissertation. K.U, Leuven (2004)
22. Hazel, B.: Impact Mechanics of Helmet Components, University of Twente, RMIT University, pp. 8–17 (2014)
23. Mills, N., Gilchrist, A.: Bicycle helmet design. IMechE 2006 vol. 220 Part L: J. Materials: Design and Applications (2006)
24. Galvanetto, U., Caserta, G., Blanco, D., Ghajari, M., Cernicchi, A.: Helmet performance and design. helmet research in the WP3 of the MYMOSA project. In: Proceedings of the 1st International Conference on Helmet Performance and Design, London, UK HPD-2013-9 (2013)
25. Tse, K.M., Lim, S.P., Tan, V.B.C., Lee, P.: A review of head injury and finite element head models. Am. J. Eng. Technol. Soc. **1**(5), 28–52 (2014)
26. Siwanto, W., Hua, Ch.: Strength analysis of human skull on high speed impact. Int. Rev. Mech. Eng. **6**(7), 1511–1512 (2012)
27. Zhang, L., Makwana, R., Sharma, S.: Brain response to promary blast wave using validated finite elements models of human head and advanced combat helmet. Neurology (2013)
28. Casert, G.D., Iannuccia, L., Galvanetto, U.: Shock absorption performance of a motorbike helmet with honeycomb reinforced liner. Compos. Struct. **93**(11), 2748–2759 (2011)

Examination of Pneumatic Bellows for the Rehabilitation of the Human Jaw

Katarzyna Koter(✉) and Witold Szulc

Lodz University of Technology, Institute of Machine Tools and Production Engineering, 90-924, Lodz, Poland
katarzyna.koter@p.lodz.pl, 168607@edu.p.lodz.pl

Abstract. This paper presents the construction of the maxillary-mandibular system and rehabilitation methods for transient limitation of jaw opening. To eliminate the drawbacks of existing rehabilitation devices, a complex device of human jaw rehabilitation has been designed. The device uses a pneumatic propulsion system to conduct jaw movement in all required planes, by a system of pneumatic actuators and custom-made pneumatic bellows. There were manufactured several bellows of different shapes and sizes to test the impact of the bellow dimensions on its work. The results of the research process on pneumatic bellows for the rehabilitation of the human jaw are presented.

Keywords: Pneumatic · Drives · Jaw rehabilitation · Soft actuators · Pneumatic propulsion

1 Introduction

From 5% to about 12% of the adults struggle with the problem of improper functioning of the temporomandibular joint [1]. There can be many reasons, ranging from degenerative changes, disease syndromes, improper functioning of the nervous system, and mental problems. The mandibular mobility disorders in patients cause several problems, ranging from everyday activities such as: eating, drinking, breathing, talking, eventually leading to improper oral hygiene, malnutrition (weight loss) and persistent pain, ending with irreversible degenerative changes related to deformation face inclusive. In the further part of the article, the medical background and the biometry of the skeleton and the muscular system of the human craniofacial system are presented to define a starting point for a proper analysis of the suitability of the rehabilitation device. Due to various diseases and the diversity of biomechanics of human organisms, reference was made to other experimental studies.

2 Construction and Biomechanics of the Jaw-Mandibular System

2.1 Construction of the Jaw-Mandibular System

The mandible is the only movable bone, and at the same time the largest and strongest bone, lying of the craniofacial skeleton. Based on Hylander, there are three distinct

D. G. Pijanowska et al. (eds.), *Biocybernetics and Biomedical Engineering – Current Trends and Challenges*, Lecture Notes in Networks and Systems 293,
https://doi.org/10.1007/978-3-030-83704-4_4

patterns of stress and deformation of the mandible. First is corporal rotation, which is responsible for an outward rotation of both relative halves of the mandible, then medial convergence which shows the change in mandibular width while functioning. The last one is dorso-ventral shear which describes the movement in the vertical plane of both mandibular halves in relation [2] (Fig. 1).

Figure 2 shows a diagram of the location and direction of the muscles that raise and lower the mandible. Mastication muscles play an important role in the biomechanism of the mandible, as their work allows a mandibular movement necessary ex. for chewing food. Following groups can be distinguished [4]:

– Lateral Pterygoid (push the jaw forwards, pulls the lower jaw to the sides);
– Temporalis Muscle (elevation the mandible, closing the mouth, pulling the jaw posteriorly);
– Medial Pterygoid Muscle (elevation the mandible, closing the mouth);
– Masseter Muscle (elevation the mandible, closing the mouth);

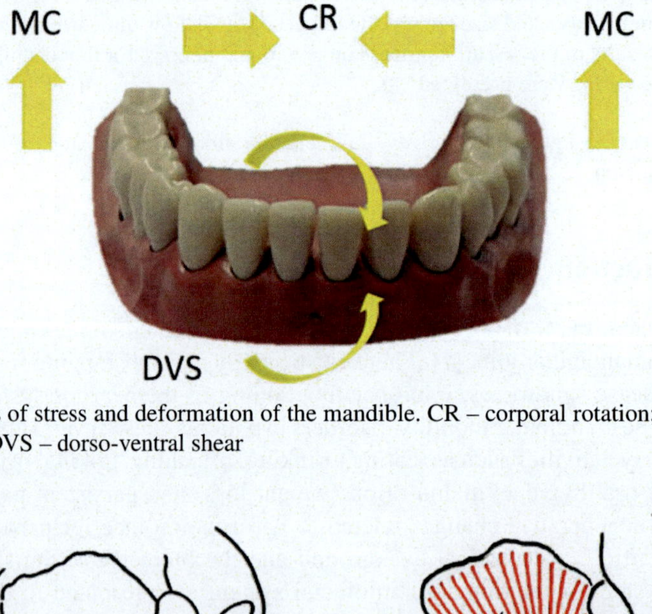

Fig. 1 Patterns of stress and deformation of the mandible. CR – corporal rotation; MC – medial convergence; DVS – dorso-ventral shear

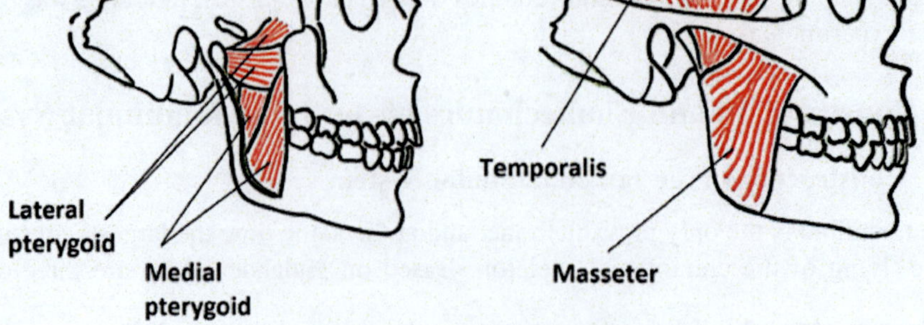

Fig. 2 Muscles of the mandible

The mandible is the arch which work can be compared to the action of the lever. The masseter muscles are attached close to the axis of rotation of the lever, and its upper attachments are not only on the jaw but also on adjacent bones and at different angles. As a result, they have a wide scope and greater force of action [4].

Research shows that in physiological conditions, the influence of muscles affects specific stress joints of the mandible. The effect of chewing inertia pressing the mandible against the maxilla causes the formation of tension and compression zones. These forces can be selected from the first premolar, part of the stretching of the alveolar part of the mandible, and impact the compressive stress on its lower edge. From the first premolar to the front, the tensile stresses decrease in favor of the torsional stress, which are the most intense at the mandible symphysis. According to analysis, the stresses in the mandible fall between 20 and 60 MPa. Greater compressive forces arise on the lower edge of the mandible near the molars and gradually decrease towards the incisors. In turn, in the alveolar part of the mandible, it is stretched under the influence of muscles; larger posterior to the first premolar and gradually narrowing down to the incisors. The closer to the symphysis of the mandible, the more a torsion force appears in its alveolar part, causing "scissor" dehiscence in the case of fractures passing in this area. This fact is important when choosing a treatment method [5, 6].

2.2 Motion of the Jaw-Mandibular System

The main 3 degrees of freedom of the mandible, allowing the lowering of the mandible, its extension/retraction and lateral right/ left movement are shown in Fig. 3.

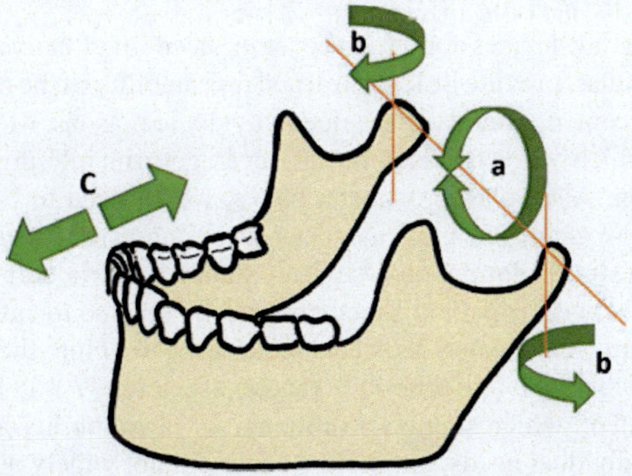

Fig. 3 The main degrees of freedom of the human mandible: **a** mouth opening and closing,**b** left and right sideways movement, **c** protrusion

The table below shows the ranges of movement of the mandible, as well as the forces necessary to achieve extreme positions [7].

Table 1. Mandible movement range and necessary forces [7]

DoFs	Movement range	Force[c]
Opening and closing the mouth	$\approx$40–70 mm[a] $\approx$28°–51°[b]	$\approx$20 N
Lateral left and right movement	$\approx \pm 4$	$\approx$25 N
Protrusion	$\approx$10 mm	$\approx$36 N

[a]Maximum distance between the incisors of the upper and lower middle incisors.
[b]Angular displacement of the lower jaw relative to the skull.
[c]Force applied to the incisal edge of the lower middle incisors.

3 Limitation of Mobility of the Temporomandibular Joint and Methods of Rehabilitation

Transient limitation of jaw opening due to reflex contracture of the muscles that raise the lower jaw is called trismus. It affects the range of mouth opening, by decreasing it from 35 to 55 m to couple millimeters only. It is usually caused by inflammation or trauma. Other causes may be tetanus, hysteria, and some CNS diseases. The treatment consists of removing the trismus cause, applying mechanotherapy and warming up the mandible. Restoring the full functionality of a jaw after mechanical injury or tooth extraction requires rehabilitation by using special tools enabling the mechanical opening of the jaw. In-home conditions, it is recommended to use corks, a wooden spoon, a spatula or a wooden wedge for exercise [8].

However, existing devices don't fully cover the needs of effective therapy. Commonly used wooden spatulas, provide little control and repeatability of the rehabilitation process and require replacement after each use due to hygienic reasons. Mechanical devices, as TheraBite [8], which work is based on the leverage principle don't ensure swift and repeatable motion. Also, plastic or metal plates, which need to be inserted inside the patient's mouth to execute jaw movement can cause damages to teeth in the longer term. One of the interesting solutions is a device equipped with a soft pneumatic actuator, which is located between a patient's teeth that can be inflated to enforce jaw movement. In the device, there were created dedicated teeth casts, to fit into the patient's mouth [7].

Additionally, there are few robotic [9] and exoskeleton [10, 11] concepts of trismus treatment equipment, which support a full range of jaw mobility and allow adaptation of a routine to individual needs, but these devices are not widely available.

To eliminate all the drawbacks of existing devices and to enlarge motion range, a complex device of human jaw rehabilitation [12] has been designed in the Institute of Machine Tools and Production Engineering of Lodz University of Technology. The device uses a pneumatic propulsion system to conduct jaw movement in all required planes. The main part of the device is a movable frame, to which a system of pneumatic actuators and custom-made pneumatic bellow is attached. Two pneumatic actuators support the additional movement of the mandible while pneumatic bellows are responsible for opening the jaw mechanism. Bellows are placed between two plates which are inserted between

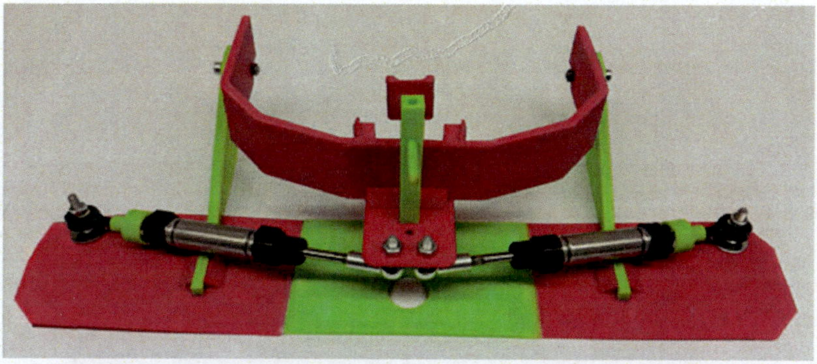

Fig. 4. The functional model of human jaw rehabilitation device

the patient's teeth. Filling of the bellows allows opening the jaw while changing the inflating pressure allows controlling the force value set on the teeth and support the smooth and repeatable motion (Fig. 4).

4 Research of Pneumatic Bellows

4.1 Pneumatic Bellows

The design of pneumatic bellows is crucial for the proper functioning of the device. The effective operation of the bellow requires adhesion to the largest possible area of the plates inserted between the teeth during rehabilitation. This will reduce the risk of teeth and enamel damage due to uneven pressure on individual teeth. The variety of sizes of human jaws means that the dimensions of the bellows must be adapted to the size of the jaw to ensure comfortable and effective rehabilitation. Also, the thickness of the actuator should be small enough to allow free application of the device equipped with pneumatic bellows into the mouth of the patient affected by the trismus.

Based on the determined approximate value of the mandible range of motion and the forces necessary to open and close the jaw the prototype of pneumatic bellows was designed. It was created with the use of composite material made of approximately 0.3 mm thick polyester fabric with a silicone layer attached. There were manufactured several bellows of different shapes and sizes to test the impact of the bellow dimensions to its work.

The picture below (Fig. 5) shows the pneumatic bellows that were tested.

4.2 Measuring Station

The measuring station (Fig. 6) was built using rigid aluminium frame (5) on which a force sensor FN3148 (1) with a measuring range up to 50 N was permanently attached on the left side, while on the right side, to the frame (5), the micrometric table (4) was attached. A handle imitating the stationary upper jaw (2) was screwed to the force sensor (1), while the mandible was simulated using the holder (3) attached to the micrometric table MX50-SC-28 (4). The handles (2) and (3) were designed using CAD software and printed on a 3D printer.

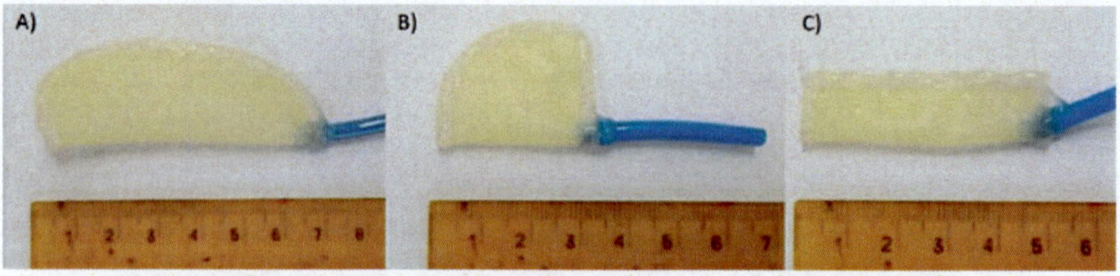

Fig. 5 Pneumatic bellows, **A** type A, **B** type B, **C** type C

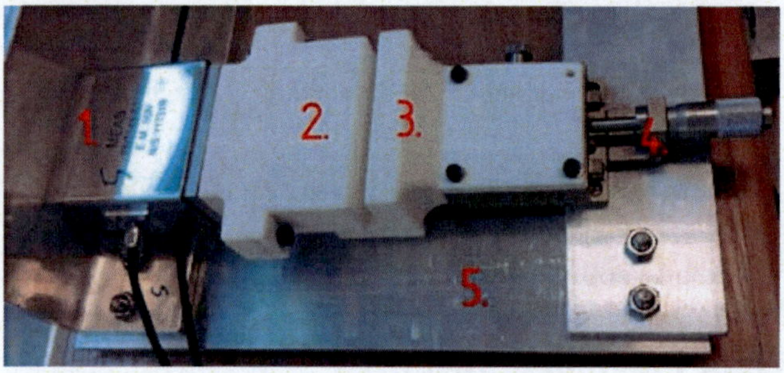

Fig. 6 Measuring station: 1) Force sensor; 2,3) Bellow holders; 4) Micrometer table; 5) Stand frame

4.3 Research

The test of the bellows was carried out on the measuring station by placing empty bellows in the gap between the jaws. Then, the bellows were filled with air at a pressure of 0.05 and 0.1 MPa with the jaws open from 4 to 16 mm (gap), with a 2 mm stroke. The jaw opening was set using a micrometer table. The response generated by the inflated pneumatic bellow was measured with a force sensor and then sent as a signal via a measurement card to a computer. The computer is equipped with software that records subsequent measurements. Figure 7 shows a block diagram of a test stand.

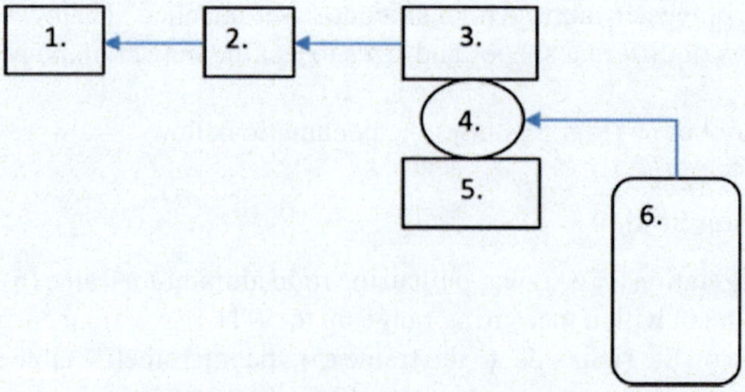

Fig. 7 Block diagram of the test stand: 1) Computer; 2) Measuring card; 3) Force sensor; 4) Pneumatic bellow; 5) Micrometer table; 6) Compressor

There were performed 3 tests for each bellow, for the whole range of pressure and width of the gap. Based on the test, there was calculated an average value of generated force. The conducted research measurements of the bellows are summarized in the form of two characteristics, in Figs. 8 and 9. They show the values of the measured force as the reaction to the impact of the air cushion, amounting to 0.05 and 0.1 MPa, depending on the set gap, for various types of bellows.

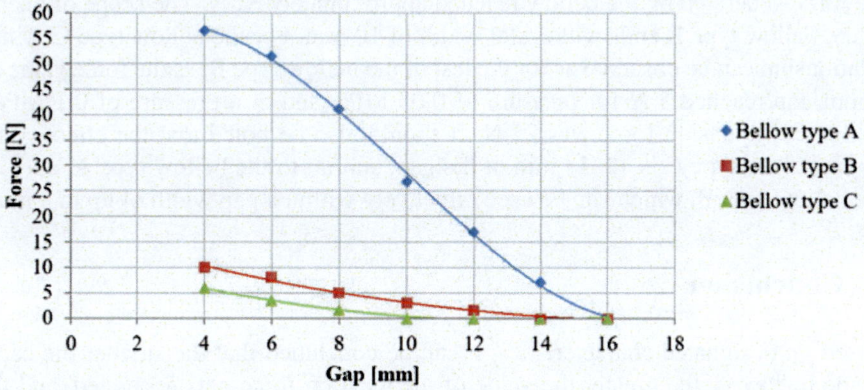

Fig. 8 Characteristics of the measured response as a function of the gap for various types of bellows, supplied with air at a pressure of 0.05 MPa

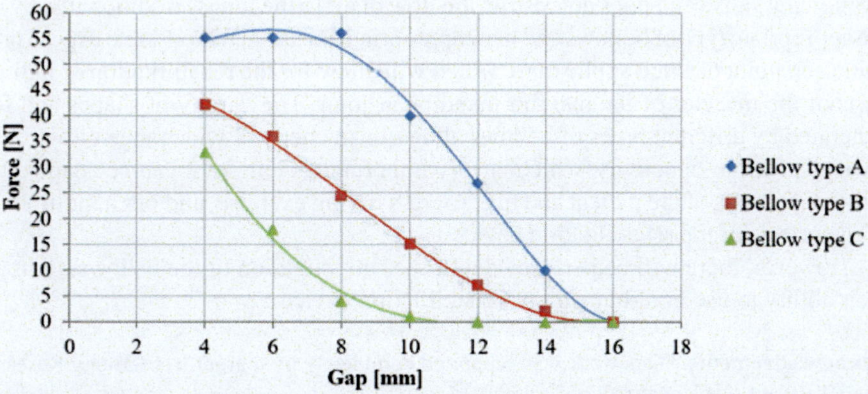

Fig. 9 Characteristics of the measured response as a function of the gap for various types of bellows, supplied with air at a pressure of 0.1 MPa

5 Results Discussion

The greatest opening force was generated by the bellow type A. It is characterized by the highest capacity of all the tested bellows. For small values of the gaps, i.e. 4–6 mm, the type A bellow reached the limit force values for the force sensor used in the measurements, regardless of the air pressure (0.05–0.1 MPa) supplying the cushion.

The bellow type A turned out to be the most effective in terms of obtained reaction values. It is especially visible with lower air pressure supplying the bellow—Fig. 8.

The type C of bellow turned out to be the weakest. Its range of motion and impact is the smallest of all bellows and practically ends with a 10 mm gap for both 0.05 and 0.1 MPa air pressure.

For the bellow type B, the force was measured at the level of 6Nfor the test gap of 4 mm and the pressure of 0.05 MPa, t, while for the same gap at the pressure of 0.1 MPa, the force generated by the bellow reached approximately 32 N. The scope of operation of the bellow type B ends with a slot width of 10 mm. For the bellow type C, a similar relationship can be observed as for the test of the bellow type B, as the force value at the 8 mm gap reached 5 N for pressure of 0.05 MPa, and at a pressure of 0.1 MPa, the generated expansion force was 25 N. It should also be noted that the effective range of the bellow type C is 12–14 mm of the gap, similar to the bellow type A, and unlike the bellow type B, which can be used effectively within a gap width of up to 8 mm.

6 Conclusion

Based on designated characteristics, it can be concluded that the smaller the capacity of the bellow is, the greater the value of the reaction force gets generated, and this is particularly visible at a small gap width.

The device placed between the teeth of the upper and lower jaws does not irritate the epithelium or mucous membranes of the oral cavity, does not cause problems with swallowing saliva and does not restrict the flow of air to the lungs. Additionally, by using the appropriate type of bellows and their appropriate location, it is possible to get a precise value and point of force application, which will allow for the rehabilitation of individual parts of the muscles of the maxillo-mandibular joint. The variety of shapes and forces generated by different types of bellows allows us to choose a wide range of options and a wide spectrum of activity, which is very important as long as it can be adapted to the individual needs of the patient and his progress during exercise, and often multiplication of the effects of the rehabilitation device.

However, there will be performed a further investigation of the bellows, to improve their ability to use in of human jaw rehabilitation device.

Acknowledgements. This work was supported completely by a grant from Young Researchers Fund 2019 at Lodz University of Technology.

References

1. Sharma, S., Gupta, D.S., Pal, U.S., Jurel, S.K.: Etiological factors of temporomandibular joint disorders. Nat. J. Maxillof. Surg. **2**(2), 116–119 (2011). https://doi.org/10.4103/0975-5950.94463
2. Hylander, W.L.: Stress and strain in the mandibular symphysis of primates: a test of competing hypotheses. Am. J. Phys. Anthropol. **64**(1), 1–46 (1984). https://doi.org/10.1002/ajpa.1330640102

3. Wong, R., Loh, J.S., Islam, I.: The Role of Finite Element Analysis in Studying Potential Failure of Mandibular Reconstruction Methods, IntechOpen (2016)
4. Lipski, M., Tomaszewska, I.M., Lipska, W., Lis, G.J., Tomaszewski, K.A.: The mandible and its foramen: anatomy, anthropology, embryology and resulting clinical implications. Folia Morphol. **72**(4), 285–292 (2013). https://doi.org/10.5603/FM.2013.0048
5. Tie, Y., Wang, D.M., Ji, T., Wang, C.T., Zhang, C.P.: Three-dimensional finite-element analysis investigating the biomechanical effects of human mandibular reconstruction with autogenous bone grafts. J. Craniomaxillofac. Surg. **34**(5), 290–298 (2006). https://doi.org/10.1016/j.jcms.2006.03.004
6. van Eijden, T.M.: Biomechanics of the mandible. Crit. Rev. Oral. Biol. Med. **11**(1), 123–136 (2000). https://doi.org/10.1177/10454411000110010101
7. Sun, Y., Lim, C.M., Tan, H.H., Ren, H.: Soft oral interventional rehabilitation robot based on low-profile soft pneumatic actuator. In: Proceedings - IEEE International Conference on Robotics and Automation, vol. 2015-June, pp. 2907–2912 (2015). https://doi.org/10.1109/ICRA.2015.7139596.
8. Lee, R., et al.: Randomised feasibility study to compare the use of Therabite® with wooden spatulas to relieve and prevent trismus in patients with cancer of the head and neck. Br. J. Oral. Maxillofac. Surg. **56**(4), 283–291 (2018). https://doi.org/10.1016/j.bjoms.2018.02.012
9. Takanobu, H., Maruyama, T., Takanishi, A., Ohtsuki, K., Ohnishi, M.: Mouth opening and closing training with 6-DOF parallel robot. In: Proceedings 2000 ICRA Millennium Conference. IEEE International Conference on Robotics and Automation. Symposia Proceedings (Cat. No.00CH37065), vol. 2, pp. 1384–1389 (2000)
10. Evans, M., Forrest, J., Tse, W., Finch, R., Xu, W.: A shoulder mounted robotic exoskeleton for rehabilitation of temporomandibular disorder via assisted motion of the jaw. In: 2016 IEEE International Workshop on Advanced Motion Control, pp. 30–37 (2016)
11. Surya, S., Ramesh, C.R., Sruthi, M., Kumar, S.: Robotic exoskeleton for rehabilitation of TMD via assisted motion of jaw. In: 2017 International Conference on Communication and Computational Technologies, pp. 172–177 (2017)
12. Koter, K., Zak, P.: A novel automated device for jaw rehabilitation. Int. J. Mech. Eng. Robot. Res. **9**(8) (2020). https://doi.org/10.18178/ijmerr.9.8.1176-1181

An Innovative Approach for a Hip Disorders Rehabilitation

Jolanta G. Zuzda[1], Jakub Kacpura[2], Jakub Dziura[2](✉), Piotr Borkowski[3], and Robert Latosiewicz[4]

[1] Institute of Management and Quality Science, Bialystok University of Technology, 16-001 Bialystok, Poland
j.zuzda@pb.edu.pl
[2] Rehabilitation Center, "Przystanek Zdrowie", 03-186, Warsaw, Poland
[3] Institute of Biomedical Engineering, Bialystok University of Technology, Bialystok, Poland
p.borkowski@pb.edu.pl
[4] Department of Physiotherapy and Rehabilitation, Medical University of Lublin, 20-059 Lublin, Poland
r.latosiewicz@umlub.pl

Abstract. Aims: With growing rate of aging population, importance of efficient and cost-effective methods to cope with musculoskeletal system disorders is essential. Almost all people at some part of life will experience problems from aforementioned system (Musculoskeletal Health in Europe. Report v5.0. http://www.eumusc.net/myUploadData/files/Musculoskeletal%20Health%20in%20Europe%20Report%20v5.pdf). It could be osteoarthritis, or rheumatoid arthritis or injuries and syndromes which result from daily living in a rushing world, where not only people overexert themselves, but from factors resulting from bad eating habits leading to obesity or malnutrition, spending many hours in one position working. There are many factors which could lead ultimately to problems with musculoskeletal diseases. Hip joint is one of the most important joints in body, as it through pelvic girdle it allows to walk, but in the same time it could be overloaded, as it transfers body weight (Kassolik K, Kurpas D, Wilk I, Uchmanowicz I, Hyży J, Andrzejewski, W (2014) The effectiveness of massage in therapy for obturator nerve dysfunction as complication of hip joint alloplasty—case report. Rehabil Nurs 39(6):311–320. https://www.ncbi.nlm.nih.gov/pubmed/23780884).

Methods: The aim of this article is the review how important is rehabilitation in hip joint problems not only to reduce present problems but also to prevent possible forthcoming injuries. Next aim is to present specially developed device to be used in addition to Hip Conditioning Program, which is special training regime and it could be great addition to physical therapy or standalone training regimen that is not only effective, but also cost-efficient.

Results: The aiding device has been designed using CAD software. The main design assumptions are: hip joint rehabilitation and co-operating muscles, active one-legged exercises without resistance or dosed resistance, two configurations (flexion–extension and internal–external rotation) with adjustable angular range. Device is simple in construction, small sized and easy to handle.

D. G. Pijanowska et al. (eds.), *Biocybernetics and Biomedical Engineering – Current Trends and Challenges*, Lecture Notes in Networks and Systems 293,
https://doi.org/10.1007/978-3-030-83704-4_5

Conclusion: Basing on the lower limb anatomy and biomechanics and the review of existing devices for rehabilitation special device with a new hip joint rehabilitation program has been developed. The device was reported in the Polish Patent Office (P. 426, 698) .

Keywords: Hip joints · Rehabilitation · Exercise therapy · Physical therapy · Osteoarthritis

1 Introduction

The one of the main causes of disabilities in Europe are disorders of musculoskeletal system [1]. These disabilities are not only causing essential decrease in people's quality of life due to pain, problems with activities of daily living or social cohabitation but also a major financial, social problem to countries' economy. Work absences, incapacity or being a lifetime pensioner cause a great burden for finance and productivity of national economy [1]. The aim of this review is to present how important are exercises in the hip joint rehabilitation process and with specially created training regime HCP, with optimal exercise load it is safe and efficient method for people who have problems with hip joints.

Hip joint diseases are one of the most significant problems in musculoskeletal system. Wrong positioning of hips and pelvic girdle can be the cause of functional dysfunction of the movement apparatus which leads to disturbance of biological functions and limit quality of life. Even slight deviation from its proper movement can lead to joint overuse and degeneration and to compensate the musculoskeletal system need to adapt so it can lead to overloading other joints and spine [1]. According to the tensegrity model created by Richard Buckminster Fuller and Kenneth Snelson in 1948 and later transformed into "biotensegrity" model by dr Stephen Lewin, problem in one place could give symptoms in other areas and even in internal organs [2, 3]. Aforementioned disturbance in biological function could be a start of many diseases of musculoskeletal system. Starting from osteoarthritis which is the second most common cause of a permanent disability. This is a degenerative disease in which joint chondroid is worn out, then it could also cause a destruction of a subchondral layer of the hip bone, capsule of the joint, synovium and tendons and it could lead to pain, impairment of range of motion. The next one is rheumatoid arthritis in which due to autoimmune inflammation of joint and surrounding tissues it causes gradually losing proper functions of joint [4]. This disease may shorten expected mean life span by 7 years [5]. But there are number of other disorders concerning hip joints - trochanteric bursitis, fractures both from traumas and due to osteoporosis, snapping hip disorder, acute or transient synovitis [6]. Advanced osteoarthritis or fractures of femur neck in the course of osteoporosis may need surgical hip replacement procedure [7]. Its aim is to improve hip function and alleviate pain. This treatment is expensive, as states in the report prepared by the International Federation of Healthcare Plans it can cost averagely, varying by geographic region, from $7,731 in Spain to even $40,364 in the United States of America [8]. It carries also some health risks like: postoperative infection, dislocation and limb length inequality, loosening of endoprosthesis and possible reoperation, impingement, nerve palsy. Chronic pain due to complications after surgery could immensely lower quality of life, and in rare cases it could lead to death [9].

In many cases this the only way to save quality of life and maintain daily living activities, but even after the procedure patient will need physical therapy for improving range of motion, muscle length and strength, decreasing muscle tone, accelerating the evacuation of postoperative edema [9].

Thanks to a progress in a medical imaging and new ways to treat hip problems modern medical society is able to restore to some extent physiological and mechanical function of hips articulations, but rehabilitation is still crucial to achieve optimal joint range of motion and pain free functioning [10].

There are many forms of physical therapy which can be used with success in this group of patients, not only before any major problems with hips but also throughout the course of disorders and after, for example, hip replacement surgery or other medical procedure to shorten the recovery time and for prevention.

Usage of physical therapy to decrease pain, stimulate muscles, improve circulation, nutrition and accelerate regeneration of the joint tissues is obtained thanks to procedures like electrotherapy, pulsed electromagnetic therapy, laser therapy, heat therapy in the form of e.g.: mudpacks, infrared therapy and many others [11, 12]. Manual therapy on the contrary is focusing mainly in restoring lost range of motion, decrease muscle tone, improving functioning of the articulation. It is based on the manual techniques which focus on the normal bone rotations within joint surfaces. Localization of restrictions and direction of it allows to manually relax the capsule of the joint using traction or roll and glide techniques [13]. Form of the manual therapy is also myofascial therapy where using many forms of massage, firm pressure it is possible to loosen muscles, alleviate pain and improve range of motion [14].

Despite the fact that aforementioned forms of physical therapy are great help to recovery process, the most important part rest on the exercise therapy—not only to strengthen often weakened or excessively tensed muscle, lengthen shortened muscle, but as form of prophylaxis and improving the quality of life [14, 15]. It is an effective and efficient form of rehabilitation and what is important for health care systems throughout the world it is cost-effective [16–19].

2 HCP—Hip Conditioning Program—Strengthening and Flexibility exercises

Movement of body parts is often based on the movement of the center of gravity of the body, usually classified in three perpendicular planes crossing that center – transverse, sagittal and frontal [20]. Rotation by definition is a circular movement around a certain point or axis. When this system is restricted or secured rotatory movement can be introduced with a use of an external force [20, 21].

In everyday tasks, such as sweeping the floor or moving objects from lower level to higher one and nearly all other kinds of daily-based activities requires rotation motion in at least one body part [22].

That kind of action is rarely performed by one joint or in a singular plane. From an anatomical point of view, most of motion activities, require multiple joints and muscle groups action, which on the other hand necessitate rotations in some joints of the body. Depending on the kind of movement, rotation will occur by a previously mentioned point or axis, where point will most likely be a single joint, (e.g. hip-joint or shoulder joint) and where axis will be a multiple joints action nearly always involving joints of the spine [22, 23]. This kind of motion is healthy from biomechanical and human physiology points of view. It allows symmetrical and stable development of the muscle and bone mass and structure, ergo making rotations crucial for all kind of daily-based activities and proper, healthy growth of the body and sustain [24, 25]. For example, when sitting at the table eating soup, the usage of a spoon requires flexion and adduction of the arm along with the rotation of the forearm and wrist connected in one chain movement [22]. Taking this into consideration, during the training of certain body parts, if possible, rotation should be introduced, making that kind of training similar to original, life-based movement [25].

Hip Conditioning Program—Strengthening and Flexibility Exercises (HCP)—introduces exercises for beginners in poor fitness and/or health condition, suffering from movement restrictions caused by motor stagnation and muscle contractures, as well as for those not suffering from any kinds of movement restrictions. Program contains exercises that most often require complex movement of all body parts, activating not only superficial and large muscle groups, but also deep and short muscles. Such activation rarely occurs in linear or isolated movements. Better activation leads to extension of muscular work, stabilization of the body and in consequence better movement and balance control during movement of the gravity center of the body. The twisting force that takes place during rotations allows for the lengthening of the muscles and tendons, thus removing or decreasing contractures and stiffness [24].

HCP introduces rotational movements, that aren't directly opposed to gravitational force, as linear ones, which inflict less stress on cartilage of articulations. Previously mentioned twisting force combined in HCP regime training can improve the structures and functions of the ligaments, tendons and joints, increasing their flexibility. Continuous training in HCP regime will contribute to proprioception, coordination, mobility and motor efficiency of the body and achieving better health status, thus decreasing joint and muscle degeneration, the risk of falls and occurrence of numerous traumatic injuries in elderly groups, while increasing fitness and health condition in all ages [24].

Some examples of rotational movements are introduced in Kabat-Kaiser method, proprioceptive methods and many others methods used in existing rehabilitation programs [17, 28]. Other than that, rotational movements can be found in some preventive-corrective exercises in scoliosis treatment and most positions in Yoga, Tai Chi, Body ART [21, 26, 27], as well as in Pilates classes, stretching and core sessions training [28, 31].

The goal of the initial stage of training is to begin the adaptive processes. Typically, this is accomplished by working out at lower intensity and with shorter duration. Signs of intensity load intolerance particularly from musculoskeletal or cardiovascular system should be carefully evaluated. The initial stage is the time to develop the habit of exercising with minimal discomfort and muscle soreness. It is also a time to allow the therapist to instruct the individuals how to exercise properly.

The point of the improvement or intermediate level stage of training is to provide a gradual increase in the overall exercise stimulus to allow further improvements in the musculoskeletal health, and it is accomplished by increasing frequency, duration or intensity. Total weekly training volume should not be increased more than by 10%. Increase duration by 5 to 10 min per session on a weekly basis and intensity through the range of 60% to 85% of aerobic capacity over months. After approximately 6 months of training, most individuals make the transition from improvement to maintenance. If the workout is interrupted, profits in the fitness regress by about 50% within 4 to 12 weeks. They are all signs like pain or discomfort suggesting that progression may be too rapid [28, 29].

The purpose of this maintenance stage or advanced level of training is the long-term maintenance of musculoskeletal fitness developed in the previous stage. It is strengthened by diversification and purposeful attempts to rotate and reduce the stresses of continued training. Diversification may be used as a means of reducing the potential for overuse injuries, particularly for participants with musculoskeletal limitations. Through alternating of local fatigue and musculoskeletal stresses across different muscles groups it is possible to reduce risk of overuse injuries and lessen signs of soreness. This stage is a time for enjoyment, surveillance, and reappraisal. It is a time for enjoying the fruits of labor by competing, engaging in the new activities, or overuse injury might continue during the maintenance phase. During this stage the individual should be reexamined, physiological or performance tested and new goals may be established [29].

The Academy of Orthopedic Surgeons (AAOS) concluded that hip conditioning program should be continued for 4 to 6 weeks, and performing the exercises two to three days a week, unless otherwise specified by physical therapist, and after that can be continued as a maintenance program for lifelong protection and health of hips and thighs [30].

3 Example of HCP Exercises in Training Regimen

HCP benefits from simplicity of rotational movements, uncomplicated procedures and no need for specific equipment. The exercises should be considered as low or medium intensity and should be performed in a gentle way with moderate physical effort required. It is worth mentioning, that even the smallest difference in body posture and motion makes a difference on the overall effect of a particular exercise.

Instruction for Exercise 1

OBJECTIVE: Mobilizes your hip joints to help preventing hip injuries that commonly affect desk workers. Strengthening of the psoas major and minor muscles, iliacus muscle, rectus abdominis muscle and rectus femoris muscle, stretching gluteus maximus, medius and minimus muscle.

BENEFITS: Improve mobility in your hip joints. Mobilizes the group of adductors and strengthening muscles which flex and rotate trunk and thus improving correction of the body posture.

DESCRIPTION:

Initial position: Lying on your back. Lower limbs bent at hip and knee joints. Pelvis in neutral position. Belly draw in. Put the upper limbs along the body, lower the shoulders, lengthen the neck, bring the chin closer to the chest. Look straight ahead (Fig. 1a).

Movement: Bend the lower limbs, raise and put them on the HCP device support (Fig. 1b). Bring the lower limbs towards the chest (Fig. 1c). Hold in this position for 1–3 s, then repeat the movement 9 to 12 times. Return to starting position.

End position: as the initial one (Fig. 1a).

Repetition: Repeat 9–12 times (basing on your particular level of fitness).

Equipment needed: Rehabilitation mat, device for HCP.

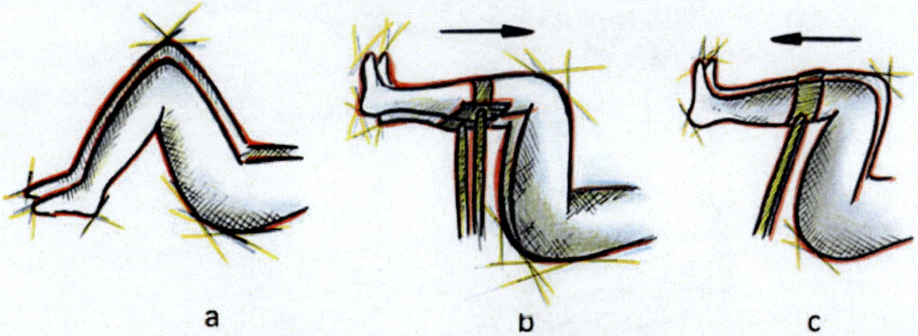

Fig. 1 Stretching of the gluteal muscles in lying on the back using the HCP device: **a** starting position, **b** lower limb flexion in the hip joint, **c** lower limb extension movement in the hip joint

4 Rehabilitation Device Designed to Use in the HCP Training Regimen

There are many devices on the market intended for exercises and rehabilitation of the joints. However, carrying out rotational exercises of the hip joints has been possible up to now by performing these exercises in suspension training mode. Devices of this type are usually large and are available mainly in specialized rehabilitation units. Therefore, we have decided to develop a portable device adapted to the requirements of the described program (HCP).

The rehabilitation device (Figs. 2 and 3) consists of 3 main units: a module for rotational exercises (1), an opposing leg support (2) and back pad (3). The mechanism (1) and the support (2) are connected to the washer (3) via knobs (4). The device allows free, multidirectional movement of the lower limb, which facilitates rotational exercises of hip joint. Exercises can be carried out in home-based program by the person exercising alone or assisted by a physiotherapist.

The exercise module has been designed primary to perform the rotational exercises of hip joint. It also enables isolated exercises of this joint in the sagittal and frontal planes, which makes it more versatile. The device was reported to the Patent Office of the Republic of Poland under the name "Device for rotational hip exercises" (P. 426, 698).

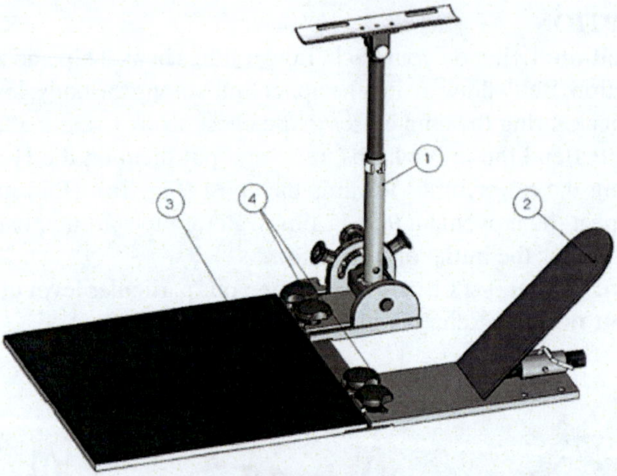

Fig. 2 3D model of HCP device

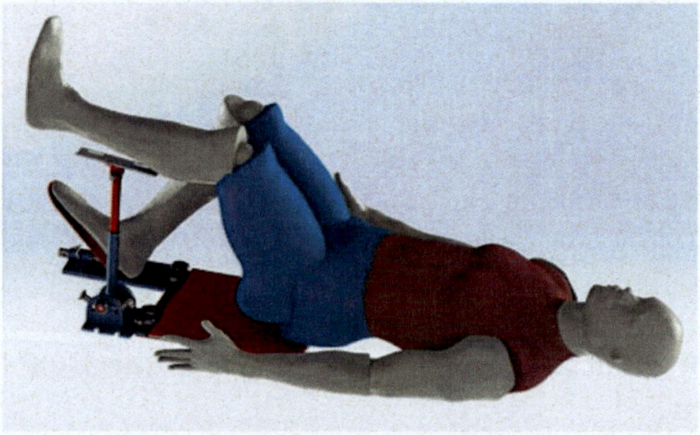

Fig. 3 3D model of HCP device

The aforementioned rehabilitation device was developed basing on rehabilitation device also created for use in HCP training by authors (Figs. 4 and 5). It also enables isolated exercises of hip joint but in sagittal and transverse plane. The device was reported in the Polish Patent Office (P.420968).

5 Discussion

Prophylaxis is a crucial part of physical therapy. Strengthening and improving flexibility of muscles and joints range of motion not only help to maintain mobility and quality of life, but also it helps to prevent falls in elderly. It is especially true in those with higher risk of fractures like people with osteoporosis, where loosening of the bone structure brings higher risk of bone traumas.

The article by James K.H. Luk et al. from 2015 show that evidence based methods based on exercises like tai-chi or home-based exercises are preferred by patients and

Fig. 4. 3D model of earlier device used in HCP training

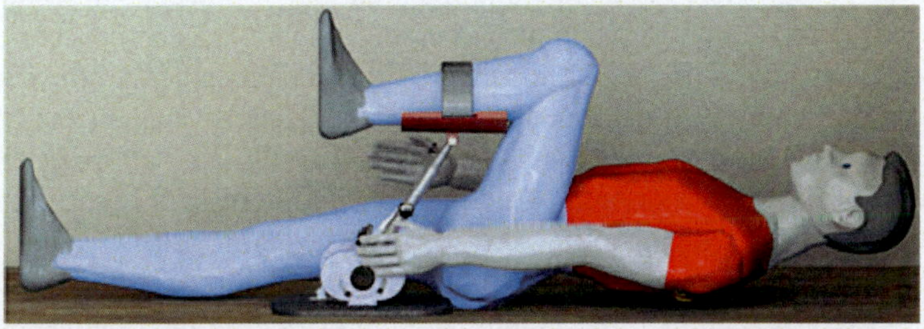

Fig. 5 Flexion of hip joint with auxiliary device help

they are efficient methods as a part of multifactorial approach to reduce risk of falls [31]. In the meta-analysis published in 2015 Hill et al. present that exercise therapy in mean of stability, balance, muscle strength and general physical activity led to a significant improvement, but it does not lead to reduce falls and bone injuries, but only in one study group [32]. It can lead to conclusion that reducing risk of falls in elderly may need more multifactorial approach, including exercise therapy but not only, also pharmacotherapy, cognitive-behavioral therapy, occupational therapy and others [32]. HCP because of its simplicity and holistic approach could be used in the hospitals and residential care home setting and be part of multifactorial approach in hip rehabilitation to reduce fall risk and increase general fitness.

HCP program might be useful also in preoperative rehabilitation, as procedures like hip arthroplasty or hip replacement. Major surgeries may have a negative impact on quality of life, mobility immediately after. Rehabilitation may prepare the musculoskeletal system for more exertion, it could also improve range of motion, prepare muscles, tendons to work in the new environment after the procedure [17, 18]. There is still a little research on this matter, as in article from 2015 by Cabilan et al.—they present information that pre-habilitation lasting more than 500 min reduced acute admissions to rehabilitation, but do not have impact on quality of life after surgical procedure [34]. It should be noted that patients after procedure should still adhere to the prescribed exercises. Thanks to developed device designed to help patients to do the HCP training regime correctly, it may be good option for people to continue exercises after discharge from a hospital.

People with osteoarthritis (OA) of the hip joint may experience pain of different levels of intensity—from mild pain to debilitating pain, problems with mobility due to decreased range of motion of a hip joint, stiffness especially in the mornings. This problem can apply to approximately one in ten individual living in industrialized countries [35]. Regular exercises can alleviate pain symptoms and improve functioning in everyday life. Exercise therapy due to nature of it, make people feel that they have impact and responsibility over their recovery process. It is very desirable approach as it could not only improve state of physical health, but also can lower stress levels and gave feeling of having some control over their bodies [34]. In 2014 in Deutsches Ärzteblatt International was published an article where authors present results of trial where group of people with hip osteoarthritis after 12 weeks of exercise therapy show significant improvement in the mean of the reduced pain and better physical functioning versus to control group and placebo group where the therapeutic intervention was placebo ultrasound therapy. Though the pain was reduced and improved mobility, there was no impact on the general-health related quality of life [35]. The authors conclude that efficacy of exercise therapy depends on how thoroughly particular exercise is conducted by patient and on adherence to training after training regimen. Beumer et al in 2016 published an article that indicates that water-based exercises or land-based present short–term efficacy in managing pain symptoms associated with hip joint osteoarthritis. The authors also postulated that more high-quality randomized control trials are needed to evaluate medium- and long-term efficacy of exercise therapy [36]. Steinhilber et al in 2017 in their publication resumed that the randomized controlled trial in which patients with OA have received 12-weeks of exercise therapy present important results. These results show significant impact of exercise therapy on hip muscle strength (evaluated as isometric peak torque of muscles responsible for hip flexion and extension and hip abduction and adduction). Not only it was effective but also feasible and easy to implement and as we mentioned earlier in this article, as patients feel that they have impact on their health adherence rate was high, up to 90% [37]. Adverse effect were mainly exercise-related pain and fatigue of intermittent nature. HCP training because of its principles and leveled difficultness may be used not only in the first stage of rehabilitation, but also it could be modified to be more advanced in later stages of physical therapy to bring more benefits and maintain fitness and control over level of pain.

The effects of HCP training are to increasing the coordination and balance of body, efficiency of the body motor and mobility functioning, strengthening muscles of the hip and lower limb region [24]. In an article by Bourne et al. from 2018 authors provide evidence-based recommendation for practitioners to understand and prescribe suitable exercise therapy for preventing hamstring injuries and disorders. Based on evaluation of many years body of work on physical therapy, especially exercise therapy they clarified that strengthening long head of biceps femoris and semimembranosus muscle help with hip extension related movements and subsequently better activation and strength of short head of biceps femoris and semitendinosus muscle play crucial role in knee flexion movement. They concluded that it is important in practice to correctly prescribe exercises having in mind what is the main problem for patient [38].

Iliotibial band syndrome connected mostly with runners may have its cause with weak hip abductors—this opinion is based on biomechanical and physiological observation. Studies on that topic have yet to conclude if it is the reason of this disorder [39]. However, in the data it is clear that factors like hip/knee coordination and functional state of lower limb joints play important role. Running is more and more popular form of recreation and incidence of this disorder is estimated to be between 5 and 14%, so it could be important that one of forms of prevention should be exercises focusing on flexibility and strength of muscles taking part in running activity [39]. On the other hand, article from 2016 present information that suggest that hip-extensor weakness forces more upright trunk positioning which can lead to over reliance on extensor muscles of a knee. To compensate that, runners exerts their knee joints which can lead not only to pain but ultimately to more serious injuries and in the long run it could raise the risk of osteoarthritis [40]. In the article by Ishøi et al. published in 2016, where group of 24 professional soccer players from U-19 units researchers add to their standard training exercises which purpose was to strengthen eccentrically hip adductors. After 8-week training regime there were significant rise in strength [41]. Authors stated that players with weaker hip adductors are more prone to injuries than players who add to their training exercises focused on strengthening aforementioned muscle parts [41]. The rehabilitation device and whole program development was accomplished thanks to dotation from European Union funds and Bialystok University of Technology [42].

6 Conclusions

1) Physical therapy, especially exercise therapy like HCP training regimen could be and should be implemented in rehabilitation in people who have problems with hip joints to strengthen, improve flexibility as a form of pre-rehabilitation and prophylaxis.
2) Rotational movements, which are normally used in daily life activities and the they are main part of HCP program allow improvement of quality and functioning in daily life for individuals with hip dysfunctions.
3) HCP program due to strengthening and increasing flexibility of muscles could be used in athletes to prevent injuries and overuse of locomotor system.

References

1. Musculoskeletal Health in Europe. Report v5.0. http://www.eumusc.net/myUploadData/files/Musculoskeletal%20Health%20in%20Europe%20Report%20v5.pdf
2. Kassolik, K., Kurpas, D., Wilk, I., Uchmanowicz, I., Hyży, J., Andrzejewski, W.: The effectiveness of massage in therapy for obturator nerve dysfunction as complication of hip joint alloplasty—case report. Rehabil Nurs. **39**(6), 311–320 (2014). https://www.ncbi.nlm.nih.gov/pubmed/23780884

3. Swanson, R.L.: Biotensegrity: a unifying theory of biological architecture with applications to osteopathic practice, education, and research-a review and analysis. J. Am. Osteopath Assoc. **113**(1), 34–52 (2013). https://pdfs.semanticscholar.org/ca4e/59b3b1bfc514812a18049bf1ab58880ea184.pdf
4. Kuryliszyn-Moskal, A., Hryniewicz, A., Dziura, J.: Balneotherapy in the treatment of rheumatoid arthritis and osteoarthritis. Forum Reumatol. **3**(3), 168–172 (2017)
5. Annegret, F., Thomas, F.: Long-term benefits of radon spa therapy in rheumatic diseases: results of the randomised, multi-centre IMuRa trial. Rheumatol. Int. **33**(11), 2839–2850 (2013). https://doi.org/10.1007/s00296-013-2819-8
6. Morelli, V., Espinoza, L.: Groin injuries and groin pain in athletes: part 2. Prim. Care **32**(1), 185–200 (2005). https://doi.org/10.1016/j.pop.2004.11.012
7. Arkusz, K., Klekiel, T., Niezgoda, N., Będziński, R.: The influence of osteoporotic bone structures of the pelvic-hip complex on stress distribution under impact load. Acta Bioeng. Biomech. **20**(1), 29–38 (2018). https://doi.org/10.5277/ABB-00882-2017-02
8. A study of cost variations for knee and hip replacement surgeries in the U.S. The Health of America Report, Blue Cross BlueShield bhi Blue Health Intelligence. 21 January 2015. Archived from the original (PDF) on 22 October 2015. Accessed 4 Oct 2015. https://www.bcbs.com/sites/default/files/file-attachments/health-of-america-report/BCBS-HealthOfAmericaReport-KneeHipReplacementSurgeries.pdf
9. Smith, T.O., Aboelmagd, T., Hing, C.B., MacGregor, A.: Does bariatric surgery prior to total hip or knee arthroplasty reduce post-operative complications and improve clinical outcomes for obese patients? Systematic review and meta-analysis. Bone Joint J. **98-B**(9), 1160–1166 (2016). https://doi.org/10.1302/0301-620X.98B9.38024
10. Bozic, J.K., Kurtz, S.M., Lau, E., Ong, K., Vail, T.P., Berry, D.J.: The epidemiology of revision total hip arthroplasty in the United States. J. Bone Joint Surg. Am. **91**(1), 128–133. https://doi.org/10.2106/JBJS.H.00155
11. Kulig, K., Burnfield, J.M.: The role of biomechanics in orthopedic and neurological rehabilitation. Acta Bioeng. Biomech. **10**(2), 3–14 (2008)
12. Bender, T., Karagülle, Z., Bálint, G.P., et al.: Hydrotherapy, balneotherapy, and spa treatment in pain management. Rheumatol Int. **25**(3), 220–224 (2005). https://doi.org/10.1007/s00296-004-0487-4
13. Hammad, S.M., Arsh, A., Iqbal, M., Khan, W., Bilal, S.A.: Comparing the effectiveness of Kaltenborn mobilization with thermotherapy versus Kaltenborn mobilization alone in patients with frozen shoulder [adhesive capsulitis]: a randomized control trial. J. Pak. Med. Assoc. **69**(10), 1421–1424 (2019). https://doi.org/10.5455/JPMA.279
14. Martínez Rodríguez, R., Galán del Río, F.: Mechanistic basis of manual therapy in myofascial injuries. Sonoelastogr. Evol. Control. J. Bodyw. Mov. Ther. **17**(2), 221–234 (2013). https://doi.org/10.1016/j.jbmt.2012.08.006
15. Hayden, J.A., Wilson, M.N., Stewart, S., Cartwright, J.L., Smith, A.O., Riley, R.D., van Tulder, M., Bendix, T., Cecchi, F., Costa, L.O.P., Dufour, N., Ferreira, M.L., Foster, N.E., Gudavalli, M.R., Hartvigsen, J., Helmhout, P., Kool, J., Koumantakis, G.A., Kovacs, F.M., Kuukkanen, T., Long, A., Macedo, L.G., Machado, L.A.C., Maher, C.G., Mehling, W., Morone, G., Peterson, T., Rasmussen-Barr, E., Ryan, C.G., Sjögren, T., Smeets, R., Staal, J.B., Unsgaard-Tøndel, M., Wajswelner, H., Yeung, E.W.: Chronic Low Back Pain IPD Meta-Analysis Group. Exercise treatment effect modifiers in persistent low back pain: an individual participant data meta-analysis of 3514 participants from 27 randomised controlled trials. Br. J. Sports Med. (2019). https://doi.org/10.1136/bjsports-2019-101205

16. Moss-Morris, R., Harrison, A.M., Safari, R., Norton, S., van der Linden, M.L., Picariello, F., Thomas, S., White, C., Mercer, T.: Which behavioural and exercise interventions targeting fatigue show the most promise in multiple sclerosis? A systematic review with narrative synthesis and meta-analysis. Behav. Res. Ther. **28**, 103464 (2019). https://doi.org/10.1016/j.brat.2019.103464
17. Krauss, I., Mueller, G., Haupt, G., et al.: Effectiveness and efficiency of an 11-week exercise intervention for patients with hip or knee osteoarthritis: a protocol for a controlled study in the context of health services research. BMC Public Health. **16**, 367 (2016). https://doi.org/10.1186/s12889-016-3030-0
18. Borkowski, P., Zuzda, J., Latosiewicz, R., Stepaniuk, K.: Program for preventing degenerative changes in the hip joint-proposal of exercise and a device. In: 20th Polish Conference on Biocybernetics and Biomedical Engineering: Abstract book. Polish Academy of Science, Polish Society of Biomedical Engineering, AGH University of Science and Technology, Kraków, p. 94 (2017)
19. Nikitovic, M., Wodchis, W.P., Krahn, M.D., Cadarette, S.M.: Direct health-care costs attributed to hip fractures among seniors: a matched cohort study. Osteoporos. Int. **24**(2), 659–669 (2013). https://doi.org/10.1007/s00198-012-2034-6
20. Dancy, P.J. (ed.) ACSM's Guidelines for Exercise Testing and Prescription, pp. 247–249. ACSM, Philadelphia, Baltimore, New York, London, Buenos Aires, Hong Kong, Sydney, Tokyo: Lippincott Williams & Wilkins (2006). ISBN-13 978-0781745062
21. Coulter, H.D.: Anatomy of Hatha Yoga: A Manual for Students, Teachers, and Practitioner's. Body and Breath, Marlboro (2010). ISBN 9780970700612
22. Stanziano, D.C., Signorile, J.F., Mow, S., Davidson, E.E., Ouslander, J.G., Roos, B.A.: The modified total body rotation test: a rapid, reliable assessment of physical function in older adults. J. Am. Geriatr. Soc. **58**(10), 1965–1969 (2010). https://doi.org/10.1111/j.1532-5415.2010.03060.x
23. Varadhan, S., Zhang, W., Zatsiorsky, V.M., Latash, M.L.: Age effects on rotational hand action. Hum. Mov. Sci. **31**(3), 502–518 (2012). https://doi.org/10.1016/j.humov.2011.07.005
24. Zuzda, J.G., Topczewska, M., Borkowski, P., Latosiewicz, R.: The influence of rotational training on muscle activity of young adults in thermographic imaging. Stud. Logic Grammar Rhetoric **56**(1), 91–105 (2018). https://doi.org/10.2478/slgr-2018-0043
25. McAtee, R.E., Charland, R.E.: Facilitated Stretching PNF stretching and strengthening made easy. Human Kinetics, Champaign (2007)
26. Steinbacher, R., Alexa, L.: Body ART. Riva Verlag (2011). ISBN 978-3-86883-053-8
27. Sanders, M.E.: On the floor: hit the surf before the turf: healthy back training. ACSMS Health Fit J. **8**(6), 30–33 (2004). https://doi.org/10.1097/00135124-200411000-00011
28. Willardson, J.M.: A periodized approach for core training. ACSMS Health Fit. J. **12**(1), 7–13 (2008). https://doi.org/10.1249/01.FIT.0000298458.22383.bd
29. ACSM: The recommended quantity and quality of exercise for developing and maintaining cardiorespiratory and muscular fitness and flexibility in healthy adults. Med. Sci. Sports Exerc. **30**(6), 975–991 (1998). https://doi.org/10.1097/00005768-199806000-00032
30. Lehmann, M., Foster, C., Keul, J.: Overtraining in endurance athletes: a brief review. Med Sci Sports Exerc. **25**(7), 854–862 (1993)
31. American Academy of Orthopaedic Surgeons, Hip Conditioning Program. https://orthoinfo.aaos.org/en/recovery/hip-conditioning-program/. Accessed 01 Dec 2019
32. Luk, J.K., Chan, T.Y., Chan, D.K.: Falls prevention in the elderly: translating evidence into practice. Hong Kong Med. J. **21**(2), 165–171 (2015). https://doi.org/10.12809/hkmj144469
33. Hill, K.D., Hunter, S.W., Batchelor, F.A., Cavalheri, V., Burton, E.: Individualized home-based exercise programs for older people to reduce falls and improve physical performance: a systematic review and meta-analysis. Maturitas **82**(1), 72–84 (2015). https://doi.org/10.1016/j.maturitas.2015.04.005

34. Cabilan, C.J., Hines, S., Munday, J.: The effectiveness of prehabilitation or preoperative exercise for surgical patients: a systematic review. JBI Database Syst. Rev. Implement Rep. **13**(1), 146–187 (2015). https://doi.org/10.11124/jbisrir-2015-1885
35. Krauß, I., Steinhilber, B., Haupt, G., Miller, R., Martus, P., Janßen, P.: Exercise therapy in hip osteoarthritis - a randomized controlled trial. Dtsch Arztebl Int. **111**(35–36), 592–599 (2014). https://doi.org/10.3238/arztebl.2014.0592
36. Beumer, L., Wong, J., Warden, S.J., Kemp, J.L., Foster, P., Crossley, K.M.: Effects of exercise and manual therapy on pain associated with hip osteoarthritis: a systematic review and meta-analysis. Br. J. Sports Med. **50**(8), 458–463 (2016). https://bjsm.bmj.com/content/50/8/458
37. Steinhilber, B., Haupt, G., Miller, R., Janssen, P., Krauss, I.: Exercise therapy in patients with hip osteoarthritis: effect on hip muscle strength and safety aspects of exercise - results of a randomized controlled trial. Mod. Rheumatol. **27**(3), 493–502 (2017). https://doi.org/10.1080/14397595.2016.1213940
38. Bourne, M.N., Timmins, R.G., Opar, D.A., et al.: An evidence-based framework for strengthening exercises to prevent hamstring injury. Sports Med. **48**(2), 251–267 (2018). https://doi.org/10.1007/s40279-017-0796-x
39. Van der Worp, M.P., van der Horst, N., de Wijer, A., Backx, F.J., Nijhuis-van der Sanden, M.W.: Iliotibial band syndrome in runners: a systematic review. Sports Med. **42**(11), 969–992 (2012). https://doi.org/10.2165/11635400-000000000-00000
40. Teng, H.L., Powers, C.M.: Hip-extensor strength, trunk posture, and use of the knee-extensor muscles during running. J. Athl. Train. **51**(7), 519–524 (2016)
41. Ishøi, L., Sørensen, C.N., Kaae, N.M., Jørgensen, L.B., Hölmich, P., Serner, A.: Large eccentric strength increase using the Copenhagen Adduction exercise in football: a randomized controlled trial. Scand. J. Med. Sci. Sports **26**(11), 1334–1342 (2016). https://doi.org/10.1111/sms.12585
42. Borkowski, P., Zuzda, J.G., Latosiewicz, R.: Program of rehabilitation and prevention of musculoskeletal disorders due to rehabilitation of the hip joint - development of exercises and support devices. Prace przedwdrożeniowe sfinansowane w ramach Inkubatora Innowacyjności+, Białystok, pp. 37–42 (2019). Accessed 03 Mar 2020. http://www.bialystokonline.pl/inkubator-innowacyjnosci-uczelnie-podsumowaly-efekty,artykul,110105,6,1.html

EEG Based Image Reconstruction Using Transformers

Kacper Ogórek(✉), Paweł Poryzała, and Paweł Strumiłło

Institute of Electronics, Lodz University of Technology, Lodz, Poland
{pawel.poryzala,pawel.strumillo}@p.lodz.pl

Abstract. In this work the possibility of creating an end-to-end system being able to reconstruct natural images based on the subject's EEG response has been evaluated. Herein, various Transformer-based model's configurations have been implemented and tested in order to find the optimal solution for the given problem, taken into account the limitations of 16-channel EEG apparatus. Pretrained BigBiGAN neural network model has been used for the purpose of image featurization that is essential for the proper EEG feature mapping. The resulting image reconstruction performance was compared between the BigBiGAN and the model that is based on an original implementation of a simple DNN autoencoder. The presented implementation and obtained results show that on a limited amount of data it was very possible to build a system that is capable of successful reconstruction of images that are similar according to quantitative measures to the original graphical stimuli.

Keywords: EEG · Deep Neural Networks · Image reconstruction · SOM · GAN · Transformers

1 Introduction

Recording of the brain activity can be a very important matter enabling deeper understanding how the brain functions. EEG can be used to record electrical activity of different brain regions. The most common application of EEG recordings is the evaluation of patient's neurological condition. EEG examination can be helpful in epilepsy therapy, sleep problems investigation, concentration distractions or numerous other psychological ailments. Additionally, at present, EEG techniques are being incorporated in brain computer interface (BCI) systems. BCI can help the disabled people to interact with computers or improve patient's psychological skills such as concentration, attention or stress control [1–3]. In order to obtain sufficient amount of information needed for some BCI assignments of brain activity examination, it is often necessary or recommended to incorporate complex devices like MR scanner or 128-channel EEG system. These instruments are very expensive, therefore, not always easily accessible.

D. G. Pijanowska et al. (eds.), *Biocybernetics and Biomedical Engineering – Current Trends and Challenges*, Lecture Notes in Networks and Systems 293,
https://doi.org/10.1007/978-3-030-83704-4_6

A system proposed by Grigory Rashkov and others [4] for image reconstruction using EEG signals data consists of several parts of which first one is an EEG feature extractor. It incorporates several machine learning (ML) techniques in order to reduce the dimensionality of the raw signal input. Afterwards features are passed into the feature mapper which converts these values into the other—latent—space. Finally, latents are used for image reconstruction using image generator network. A similar, yet, more sophisticated approach was applied in research on fMRI image reconstruction by Guohua Shen and others [5]. Similarly, Highly-dimensional features of fMRI signals were compressed using feature mapper. However, these features were mapped directly into the space of image features returned by the given image encoder. Additionally, proposed method of image reconstruction is very extraordinary. Instead of incorporating a network capable of generating images out of encoded image features, in this work a standalone generative adversarial networks (GAN) image generator was used. Because of the fact that this GAN's input features are incompatible with deep neural network (DNN) outputs used for images featurization, GAN's inputs, starting with random values, were iteratively optimized in order to generate image that triggers the most alike response of the DNN.

As it was approached in previous attempts to recreate images based the brain activity registrations, brain activity data will not be translated into the images directly. EEG features will be firstly created based on the raw signal data and only after these features will be mapped into the image features latent space from which image reconstruction should be performed. Furthermore, in this research possibility of reducing input data dimensionality using tokenization will be evaluated. In this way the risk of overtraining the whole system using relatively small dataset can be minimized. In this study we demonstrate that we can reconstruct different types of images presented to the individuals from EEG recordings. Our work was inspired be earlier studies reported in literature [4, 5]. Here, a possibility of low cost 16-channel EEG system exertion for complex task of the observed image reconstruction will be investigated.

2 Materials and methods

2.1 Image Dataset

In order to acquire a balanced set of various images, the COCO dataset [6] were chosen. Because of the limited computing capabilities, the "2017 Val images" was employed for further operations. This image dataset is relatively small and consists of only 5000 images of different categories. To equilibrate the small amount of data collected during the EEG tests, 8 image categories have been introduced. For each category images has been chosen manually out of all 5000 ones. Samples of these sorted images along with the categories to which they belong are shown in Table 1.

Table 1 Samples of the categorized pictures used for the EEG tests

Animal	Food	Nature	Person	Plants	Room	Town	Vehicle

2.2 EEG Registration

Subjects

The EEG registration sessions were performed on a group of 10 healthy volunteers including 4 males and 6 females. The subjects' age was in the range from 21 to 24 years and a mean of 22 years. Prior to the session every subject was informed about the scope of the test as well as they signed the declaration of consent.

Registration Setup

EEG registration sessions took place in the broad room in which a proper test station had been prepared. This test station consisted of a 23″ LCD screen placed on a desk in front of an ergonomic chair on which subject was placed during the examination. A 16-channel g.tec EEG system was used for signal registrations. The EEG system was placed on the back of the chair. The LCD screen was connected to the server using a long HDMI cable. The experimental bench is presented in Fig. 1.

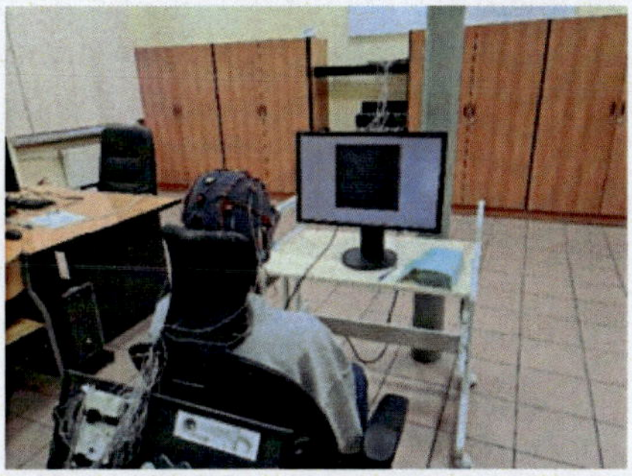

Fig. 1 Image presenting one of the subjects during the test process

Registration Procedure
To simplify and standardize EEG signal data acquisition, a specialized system provided with intuitive user interface was built. The role of the system was to synchronize EEG signal recordings in which graphical stimuli were resented to the volunteers. During each session a subject was presented with a set of approx. 40 randomly selected pictures. Every picture presented on the LCD screen was exposed for 4 s and then it was followed by a 2 s break when blank image would be displayed.

2.3 Implemented Models

Image Generation
Alongside with BigBiGAN, a custom autoencoder (CA) was implemented for image generation in the reconstruction process. While GAN architecture provides better results for image feature distribution, autoencoders are simpler to train using a limited computing power. Furthermore, because of the same reason, CA resolution was limited to 96 × 96. BigBiGAN generator creates images of a 128 × 128 resolution. In this work these different approaches to the problem of image reconstruction from EEGs was tested. Because of the limited size of the training data and available computing power, the transfer learning of BigBiGAN generator was not undertaken. Instead, an alternative method was investigated.

The method applied for mapping EEG features into BigBiGAN encoded latents was based on transformer network. The other method was aimed to boost the final reconstruction quality. For this purpose a new BigBiGAN compliant CA (BCA) model was trained. Because of the different dimensionality of the BigBiGAN's generator and the CA's decoder inputs, a feed forward network (FFN) of input size compatible with BigBiGAN's generator and output size compliant with CA's decoder was stacked together with the CA's decoder itself. This model was trained to recreate the images from BigBiGAN latents, so that, it would serve a similar role as BigBiGAN generator while being smaller and easier to train. This model was unable to perform as well as the original generator, however, it was stacked at the end of the pretrained EEG to BigBiGAN's latents mapper in order to increase the final accuracy of reconstruction.

CA's architecture is presented in Fig. 2. It is much less complex than the ResNet50/RevNet50 network for the purpose of simpler training process. This network was trained on the set of approx. 5000 natural images from the [6] dataset. Because of the limited network capabilities the reconstruction quality is not excellent, yet it is sufficient for recognizing shapes and colors. In order to simplify the learning process, input images of 96 × 96 pixels were blurred using a 5 × 5 Gaussian kernel. Samples recreated using this network are presented in Table 2.

Since the proposed CA's architecture is relatively shallow compared to the BigBiGAN network and because the image dataset is not large enough, features encoded by the CA are related mainly to image layout and colors. On the other hand features encoded by the BigBiGAN's encoder (ResNet50) can be associated with the abstracts like "dog", "sunny weather" or "ocean". Samples recreated using this network are presented in Table 3.

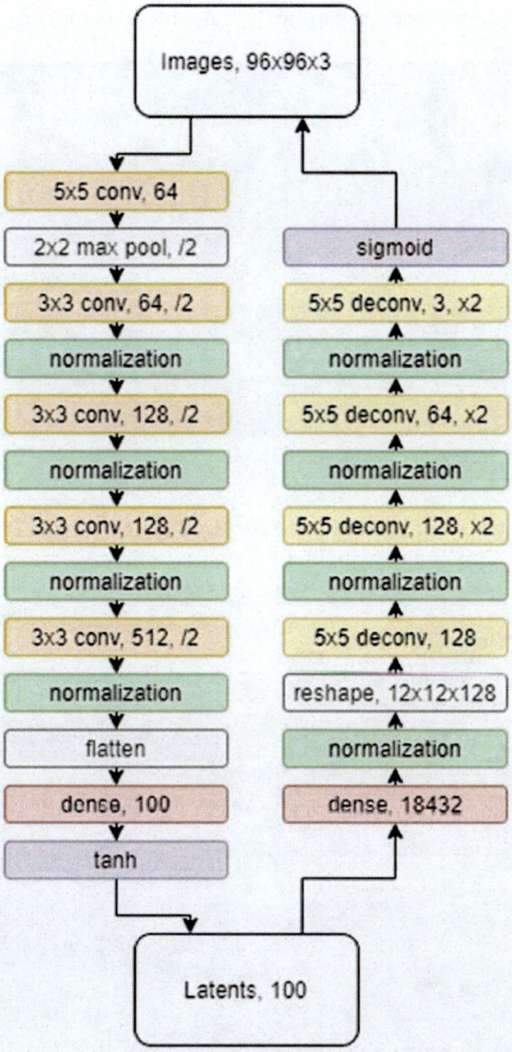

Fig. 2 The implemented CA model architecture

Tokenizer

For the purpose of EEG feature data distribution smoothening, a SOM network based tokenization system was employed. Tokenization scheme is shown in Fig. 3. Tokenizer had to be previously trained on a whole training dataset of flattened features. A trained SOM network is performing an N-dim vector features mapping onto a 2D plane. In order to tokenize a single feature, most matching node's coordinates are to be assigned for an input value.

Table 2 Sample images reconstructed using the CA—original (top) and reconstructions (bottom)

Table 3 Sample images reconstructed using the BigBiGAN—original (top) and reconstructions (bottom)

Transformer
The model introduced for the purpose of EEG features to image features mapping was a transformer network. Its architecture is very similar to the encoder structure proposed in [7] and is shown in Fig. 4.

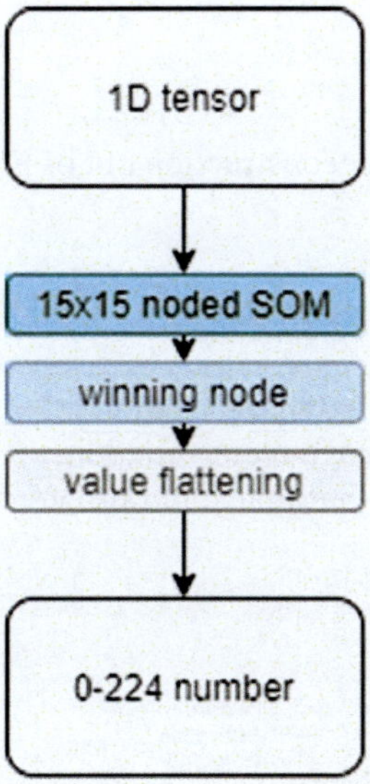

Fig. 3 The implemented tokenizer architecture

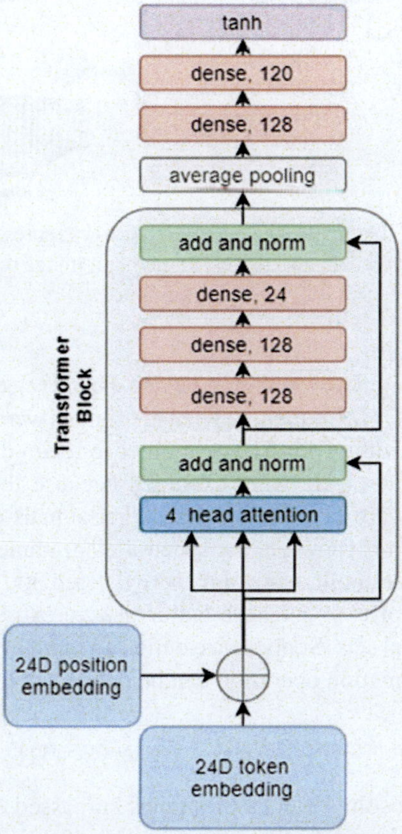

Fig. 4 The implemented single-layered transformer architecture

2.4 Analysis

Reconstruction System

The system created for image reconstruction out of EEG signals (Fig. 5) consists of the 3 main elements:

- *EEG feature extractor* converts plain EEG signals into normalized over the whole training set features.
- *Transformer* converts extracted signal features into latent space vectors which contains information sufficient for image reconstruction.
- *Image generator* creates output image out of latent space vectors obtained using the ML model.

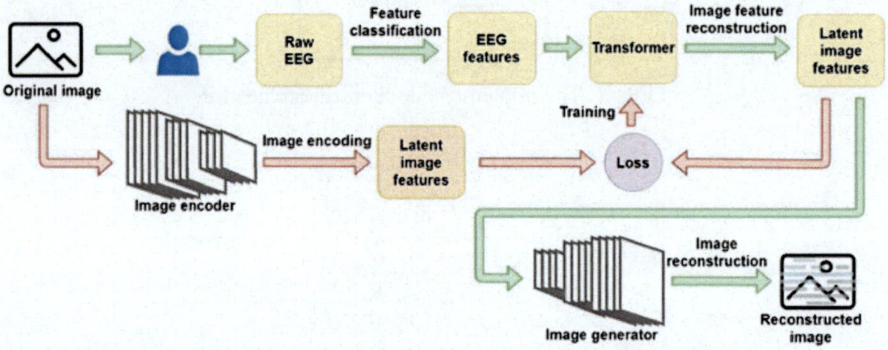

Fig. 5 General scheme of the proposed system. Red arrows represent the learning process of the regression model and the green arrows represent the main image reconstruction process

EEG Feature Extraction

Initially, the registered signals were filtered using a band rejection filter to reduce power-line contaminations of 50 Hz frequency. Then the signals were filtered again using a band pass filter with a bandwidth of 11–51 Hz in order to retain data of interest only. All the frequencies smaller than 11 Hz were excluded because the alpha wave signals in a number of recordings were far stronger than other signals of interest. An example of 10 Hz signal domination is shown in the spectral diagrammed in Fig. 6.

EEG feature sets were built using signals registered 300 ms after the start of image exposure that lasted approx. 4 s. These features were extracted from the 12-element-ICA transformed original sets. Suppose we define an input as x and the ICA transformed output as s, the transformation operation can be defined as follows:

$$(x_1, \ldots, x_{16})^T \xrightarrow{ICA} (s_1, \ldots s_{12})^T \quad (1)$$

Subsets in which absolute values of signals surpassed a standard-deviation-based threshold were excluded from further analysis. The recorded signals were stored in the subsets and then were formed into the overlapping data blocks that were transformed

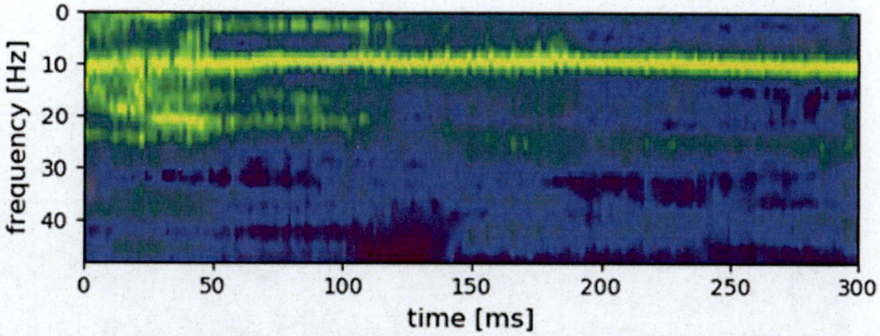

Fig. 6 A normalized sample of the EEG signal spectrum. In this example 10 Hz alpha wave is very apparent due to the subject's sleepiness

into the proper feature dataset. An i-th block b of length l derived from n-th dataset can be defied as:

$$b_n^i = s_n[i : i + l] \tag{2}$$

For each signal in each block its absolute value of the Discrete Fourier Transform (DFT) was calculated. Afterwards, every DFT was smoothed out using a low-pass filter f with a sliding window of 6 Hz defined as:

$$b_{transformed} = (f(DFT(b_1))), \ldots, f(DFT(b_{12})))^T \tag{3}$$

Subsequently, every component was normalized to the range of <0,1> . These signals were then truncated from 11 to 51 Hz. The resulting data was stacked to create a single 3D tensor B for every n-element subset as defined below:

$$B = (b_1, \ldots, b_n)^T \tag{4}$$

If transposed, this tensor would be composed of layers of spectrograms for each of the 12 signals. These spectrograms were median-blurred to reduce the noise and then normalized to a fixed mean as well as maximum values. The resulting spectrograms were timewise-flattened and portioned to be classified by the SOM network.

3 Results

In total, 2 sets of results were obtained. Image feature results obtained with the trained model were used for generating images using the BigBiGAN's generator and the BCA network. Samples of reconstructed images are presented in Table 4.

Pictures in Table 4 represent reference samples as well as images reconstructed by means of different methods. First two columns of this table present reference images of which the first one consists of original pictures presented during the EEG test sessions and the latter one consists of auto-reconstructed images from the latent space vectors which were used for training the transformer model. For this reason, the original pictures should be perceived as the reference for output images obtained using the CA's decoder

Table 4 Samples of images reconstructed using proposed system

References		Reconstruction	
Original	BigBiGAN auto-encoded	Custom Autoencoder	BigBiGAN
			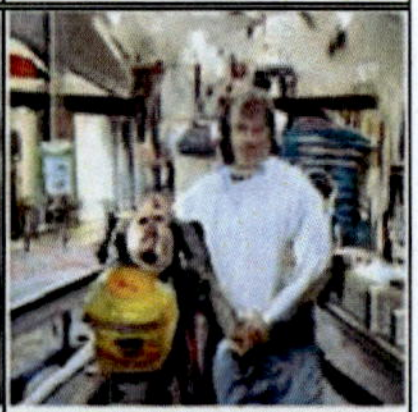

network and auto-encoded references should be recognized as references for images produced using the BigBiGAN's generator. The remaining two columns of this table present images reconstructed using the proposed system. Images in the first of these columns were obtained using CA decoder and images in the second one were created using the BigBiGAN generator.

During the network training process image reconstruction quality was assessed. Metrics incorporated for this purpose were the root mean square error (RMSE) of images and feature similarity index (FSIM) as proposed in [8]. Images reconstructed using the CA decoder network were compared with the original ones. However, image reconstructed using the BigBiGAN generator network were compared with images auto-reconstructed using the same network. Results of these comparisons are presented in Fig. 7.

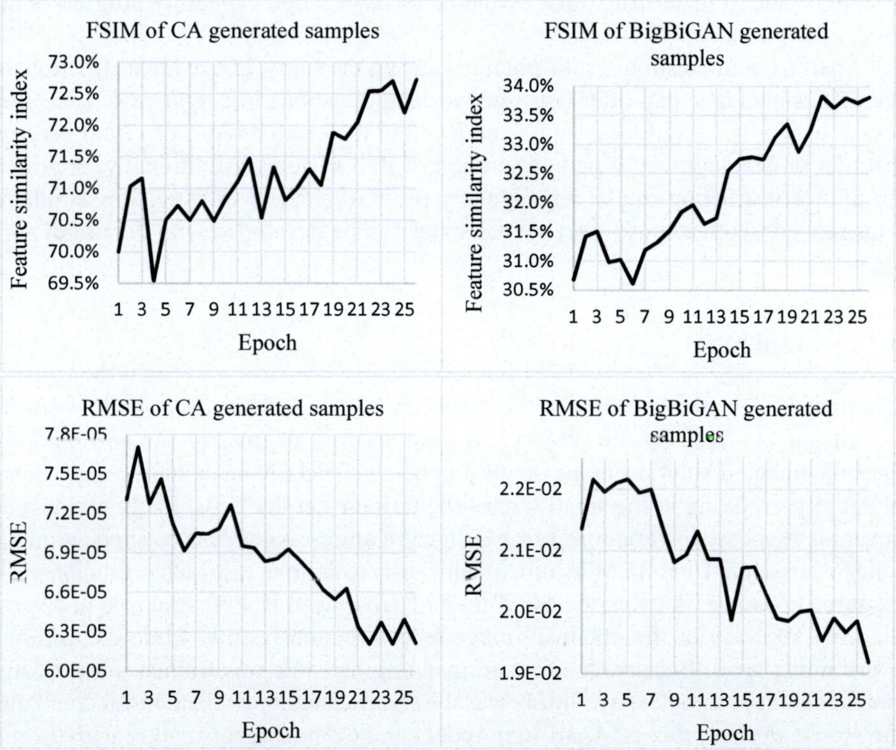

Fig. 7 Reconstruction quality metrics obtained during the training process

4 Discussion

In this study an attempt to reconstruct images presented to a subject form 16 channel EEG recordings was reported. In most cases it was possible to recreate distinguishable images that could be matched to the original images. Note that the BigBiGAN reconstructions are in most cases very similar to the original input pictures but in other cases these reconstructions are not quite accurate. As a result, some of the reconstructed images are related to the BigBiGAN auto-encoded images but do not resemble the original pictures. On the other hand images recreated using the BCA network seem to be well-fitted to the original images, yet they are obscure and blurred.

An interesting observation is that image reconstruction quality was progressing alongside with the training process. It worth noting that, metrics based reconstruction quality indices are noticeably better for the BCA-generated images. However, the

BigBiGAN-reconstructed images are of better overall quality. This can be explained by essential differences between the complexity of these two systems. The BCA decoder network was able to recreate more accurately shapes, image layout and brightness while the BigBiGAN generator network was able to preserve the overall meaning of the picture—which is evidently much more difficult to compare and does not imply any explicit structural similarity. Nonetheless, in both of these cases image reconstruction quality was constantly improving during the training process—this indicates that the proposed system is able to reconstruct images and its reconstruction capability progresses in the proper direction.

Apart from interesting results obtained during the study, image reconstruction using the EEG signal data can offer concrete world applications in cognitive science area of research. Similar techniques can be employed in the examinations of visual perception disorder or development of new generation of BCI systems. Additionally, as graphical stimuli reconstruction can be regarded as a proof of concept investigation, similar systems related to other senses and mental states can be developed with the use of similar approach.

5 Conclusions

The presented method is not perfect, however, it managed to successfully reconstruct the images observed by the subjects, to a satisfactory degree. The main limitation of the implemented solution is the small number of EEG channels and limited quantity of subjects resulting in the small size of the training set that most likely was unable to cover all the possible variations of EEG signal features. Nonetheless, application of the transformer model and SOM feature tokenization technique resulted in valuable results. Features tokenization using the SOM might have played an essential role in achieving the high accuracy of the obtained image reconstruction results, while controlling the overtraining level. Because the feature mapping network was trained using relatively small amount of samples, overfitting was the significant threat taken into account during the system design process. Apart from reducing the data dimensionality with the use of SOM network also resulted in smoothening variations in the distribution of data features. This way the risk of overfitting was significantly reduced. We hope, that using an EEG apparatus of higher number of channels and data of larger volume it would be very possible to achieve superior and more exact image reconstruction results.

References

1. Britton, J.W., Frey, L.C., Hopp, J.L. et al.: Introduction. In: St. Louis, E.K., Frey, L.C. (eds.) Electroencephalography (EEG): An Introductory Text and Atlas of Normal and Abnormal Findings in Adults, Children, and Infants. American Epilepsy Society, Chicago (2016). https://www.ncbi.nlm.nih.gov/books/NBK390346/. Accessed 09 Feb 2021
2. Farnsworth, B.: What is EEG (Electroencephalography) and How Does it Work? (2020). https://imotions.com/blog/what-is-eeg/. Accessed 09 Feb 2021
3. Graetzer, D.G.: Electroencephalography (EEG). Magill's Medical Guide (Online Edition), p. 2 (2020). http://search.ebscohost.com/login.aspx?direct=true&db=ers&AN=87690502&lang=pl&site=eds-live. Accessed 09 Feb 2021

4. Rashkov, G., Bobe, A., Fastovets, D., Komarova, M.: Natural image reconstruction from brain waves: a novel visual BCI system with native feedback. bioRxiv 787101 (2019). https://doi.org/10.1101/787101
5. Shen, G., Dwivedi, K., Majima, K., Horikawa, T., Kamitani, Y.: End-to-end deep image reconstruction from human brain activity. Front. Comput. Neurosci. **13**, 210 (2019). https://doi.org/10.3389/fncom.2019.00021
6. COCO dataset website. https://cocodataset.org/. Accessed 09 Feb 2021
7. Vaswani, A., Shazeer, N., Parmar N., Uszkoreitm, J., Jones, L., Gomez, A. N., Kaiser, Ł., Polosukhin, I.. Attention Is All You Need. arXiv:1706.03762 (2017). https://arxiv.org/pdf/1706.03762.pdf. Accessed 09 Feb 2021
8. Zhang, L., Zhang, L., Moub, X., Zhang, D.: FSIM: a feature similarity index for image quality assessment. IEEE Trans. Image Process. **20**(8), 2378–2386 (2011). https://doi.org/10.1109/TIP.2011.2109730

The Stochastic Mathematical Model Predicts Angio-Therapy Could Delay the Emergence of Metastases in Lung Cancer

Emilia Kozłowska(✉) and Andrzej Świerniak

Department of Systems Biology and Engineering, Silesian University of Technology, Gliwice, Poland
{emilia.kozlowska,andrzej.swierniak}@polsl.pl

Abstract. Lung cancer is the most common type of cancer across the world. There are known two main types of lung cancer non-small cell lung (NSCLC) and small cell one (SCLC). Here, we focus on NSCLC that is known to disseminate to distant organs such as the brain or bones. What is more, the death of a patient with this type of cancer is caused usually by metastases. It justifies tackling the problem of the emergence of metastases in NSLC in this paper. Here, we investigated which mathematical parameters affect the emergence of metastases using the mathematical oncology approach. We developed a mathematical model of lung cancer growth and metastasis dissemination by taking into account stochasticity in the process. We incorporated into the model two paths of metastasis dissemination through lymphatic and blood vessels. We performed a global sensitivity analysis to identify parameters that affect the model output (metastasis-free survival) the most. We discover that four model parameters have dominating effect on the output and two of them are controllable - the growth rate of primary tumor cells (through the administration of chemotherapeutic agents) and carrying capacity (through the administration of angio-therapy). In conclusion, a mathematical oncology approach could be applied to investigate which treatment modalities in cancer could delay the emergence of metastases.

Keywords: Stochastic differential equations · Lung cancer · Metastases · Angio-therapy

1 Introduction

1.1 Lung Cancer

Lung cancer is a leading cause of cancer death and the most common type of cancer diagnosed across the world [1]. In 2020, in Poland, 29,509 new cases of lung cancers were diagnosed. There are known two main subtypes of lung cancer non-small cell (NSCLC) and small cell (SCLC) ones, where the NSCLC constitutes 85% of all cases [2]. In addition, NSCLC could be divided into Adenocarcinomas, Squamous Cell Carcinomas, and Large Cell carcinomas, where Adenocarcinomas are the most frequent subtype of

D. G. Pijanowska et al. (eds.), *Biocybernetics and Biomedical Engineering – Current Trends and Challenges*, Lecture Notes in Networks and Systems 293,
https://doi.org/10.1007/978-3-030-83704-4_7

NSCLC [3]. High incidence and high mortality of lung cancer justify the choice of this type of cancer in the study.

1.2 Metastases of Lung Cancer

Approximately 60% of patients diagnosed with NSCLC have tumors spread to distant organs [4]. This leads to incurable cancer. Metastases are not only correlating with poor prognosis but are also associated with poor patients' performance [5].

Lung cancer metastases disseminate via two routes: hematogenous and lymphatic one [6]. It is known that metastases via the lymphatic route take longer until distant metastases are set whereas spread via blood vessels is faster. Lung carcinomas usually disseminate to the brain, bones, and adrenal glands [7–9].

The basic stages of metastasis involve i) the detachment of tumor cells from the primary tumor, ii) invasion of surrounding tissues and the basal lamina, iii) transport into the bloodstream, IV) survival and transport through lymphatic vessels or bloodstream, and V) migration, arrest, and extravasation at a distal site and formation of a metastatic lesion.

Various strategies to delay the emergence of metastases of lung cancer have been already suggested such as the application of tyrosine kinase inhibitors (TKI) [10]. However, until now there is no available treatment for lung cancer metastases which are usually resistant to most of the treatment types.

1.3 Treatment of Lung Cancer

Treatment of NSCLC depends on the stage, the histology, and the patient's condition and includes surgery, radiotherapy, chemotherapy, immunotherapy, and targeted treatment [11]. Those treatment modalities are administered alone or (more often) in combinations.

Surgery is the primary treatment type for the treatment of early-stage NSCLC. However, for stages, II-IIIA adjuvant platinum-based chemotherapy is also recommended. So far, targeted therapies have not improved overall survival in early-stage tumors [11]. Advanced-stage NSCLC, however, is not resectable and thus treated with chemotherapy combined with radiotherapy or immunotherapy. The targeted treatment has improved outcomes in a significant number of patients with advanced NSCLC. Several actionable mutations have been identified such as those in EGFR, ALK, ROS1, or BRAF among others [12].

2 Mathematical Model

We developed the model of tumor growth and metastasis dissemination in lung cancer as a stochastic one using stochastic differential equations (SDEs). We constructed the mathematical model as a three-compartmental one with the following compartments: primary tumor, lymph node, and distant metastases. We consider two routes of lung cancer dissemination hematological and lymphatic. The first route goes from cells without the metastatic potential to distant metastases through cells with metastatic potential. The second route goes from the primary tumor through lymph nodes to distant organs.

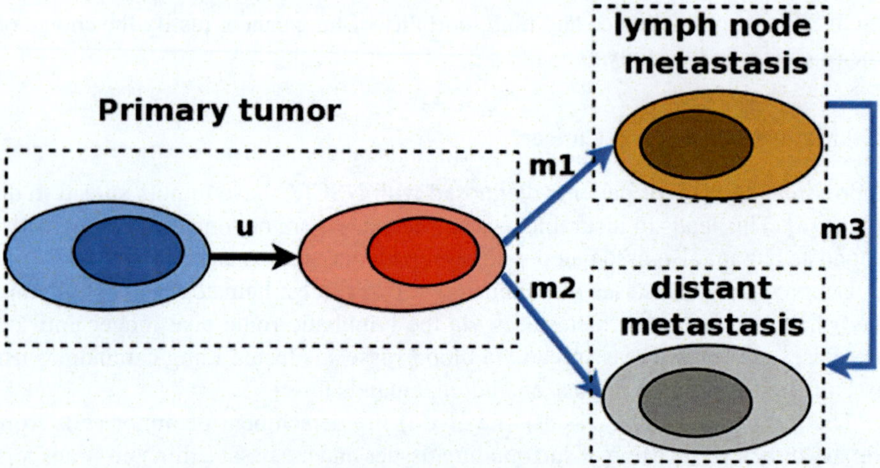

Fig. 1 Mathematical model structure

As shown in Fig. 1, the primary tumor consists of two types of cells without metastatic potential (marked with blue color) and with metastatic potential (marked with red color). The transition between them is through mutation during cell division (with mutation rate equal to u). Both types of cells grow according to Gompertzian growth dynamics with an initial growth rate equal to λ and carrying capacity equal to K.

From primary tumor, cells could disseminate to two places: lymph nodes or distant places such as the brain, bones, or adrenal glands. In addition, from a lymph node, tumor cells could disseminate to distant organs. The same as for primary tumor, cells in both compartments grow according to Gompertzian dynamics.

Taken together, the model is given with the following set of stochastic differential equations:

$$\dot{P} = (1-u) \cdot \lambda_1 \cdot P \cdot log\left(\frac{K}{P+M}\right) + \delta \cdot P \cdot W(t) \tag{1}$$

$$\dot{M} = (1-m1-m2) \cdot \lambda_1 \cdot M \cdot log\left(\frac{K}{P+M}\right) + u \cdot \lambda_1 \cdot P \cdot log\left(\frac{K}{P+M}\right) + \delta \cdot M \cdot W(t) \tag{2}$$

$$\dot{L} = (1-m3) \cdot \lambda_2 \cdot L \cdot log\left(\frac{K}{L}\right) + m1 \cdot \lambda_1 \cdot P \cdot log\left(\frac{K}{P+M}\right) + \delta \cdot L \cdot W(t) \tag{3}$$

$$\dot{D} = \lambda_3 \cdot D \cdot log\left(\frac{K}{D}\right) + m2 \cdot \lambda_1 \cdot M \cdot log\left(\frac{K}{P+M}\right) + \mathrm{m3} \cdot \lambda_2 \cdot L \cdot log\left(\frac{K}{L}\right) + \delta \cdot D \cdot W(t) \tag{4}$$

where P, M, L, D are the number of cancer cells: without metastatic potential, with metastatic potential, in lymph nodes, and distant organs. Stochasticity into the model is included by the addition of white noise represented by parameters $W(t)$ and δ.

The proposed model is the extension of the model proposed by M. Bilous et al. 2019 [13]. M. Bilous developed the model of early tumor progression of NSCLC to the advanced stage where metastases are observed. Thus, their model is predicting when metastasis will appear. The extensions of the model are the following i) inclusion of two types of dissemination: through lymph and blood vessels, ii) inclusion of tumor heterogeneity by modeling the growth of tumor cells with and without metastatic potential, and iii) addition of stochasticity to the model.

3 Global Sensitivity Analysis

To check which mathematical model parameters, affect the emergence of metastases we performed a global sensitivity analysis (GSA) which is a mathematical method for quantification of the importance of each parameter in the mathematical model in response to the output of choice.

Here, as an output, we chose metastasis-free survival (MFS) which is a time interval from the cancer diagnosis until the distant metastases are detected. By detection, we mean radiological detection which can find metastases $\geq$1 cm^3. So, when the number of cells in the distant metastasis compartment is equal to 10^9 cells (which is about 1 cm^3) the simulations are stopped, and the time of simulations is our MFS.

The GSA was performed as follows. For a given parameter, the parameters are sampled using Latin hypercube sampling (LHS) from a range: $P \pm 10\%$, where P is the nominal value of a parameter. Next, for each parameter value, 1,000 simulations are performed, and the mean value of MFS is computed for each parameter. After simulations, the values of the parameters as well as MFS are normalized. In the last step, linear regression is performed where our Y is MFS and X is P. The sensitivity coefficient is the slope estimated from linear regression which is then plotted in Fig. 3.

4 Results

4.1 Tumor Growth Dynamics

We first analyzed the dynamics of the model to understand the behavior of the model. Firstly, we plotted 1,000 time-courses of P, M, L, D (see Fig. 2). The simulations are performed from the following initial conditions:

$$P = (1 - \alpha) \cdot TumorSizeAtDiagnosis$$
$$M = \alpha \cdot TumorSizeAtDiagnosis$$
$$L = 0$$
$$D = 0,$$

where α is a fraction of cells with a metastatic potential inside the primary tumor. We assume for sake of simplicity that the nominal value of α is equal to 0.01.

As we can see in Fig. 2, each simulation represents one patient. Cancer cells in the primary tumor grow according to Gompertzian dynamics with no treatment perturbation. The dynamic of cells in the lymph node and distant metastases compartment, however, is biphasic. The first phase is when no metastases are observed and $L = D = 0$. In Fig. 3, this time is until 1000 [days]. In the second phase, the number of cells in the lymph node and distant metastases compartment grows according to Gompertzian dynamics.

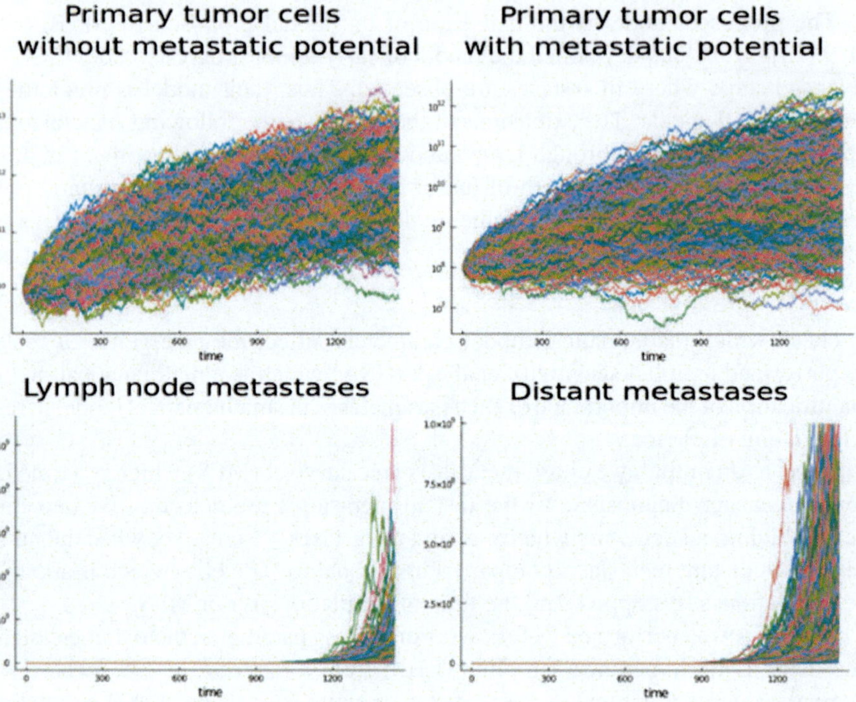

Fig. 2 Tumor growth dynamics for nominal parameter values

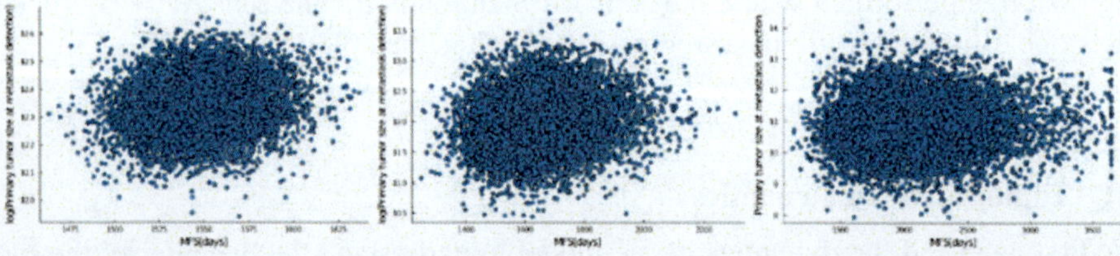

Fig. 3 Tumor burden (# of cells on a logarithmic scale) at the end of simulations as a function of metastatic-free survival

4.2 Effect of Stochasticity on the Model Behavior

Next, we performed simulations for various levels of noise which is represented with parameter σ. Thus, for three values of $\sigma = \{0.01, 0.05, 0.1\}$ we performed 10,000 simulations and plotted primary tumor size (# of cells) at the end of simulations as a function of MFS (in days).

Interestingly, with increased noise, primary tumor burden at the time of metastasis detection correlates less with metastatic-free survival. Based on Fig. 3, we can see that tumor burden does not correlate with MFS. It means, that we cannot predict solely on tumor burden if patient metastasized.

4.3 Global Sensitivity Analysis Predicts That Carrying Capacity Is the Most Sensitive Parameter in Response to Metastasis-Free Survival

To understand, which model parameters affect the model output - metastasis-free survival, we performed a global sensitivity analysis using Latin hypercube sampling (LHS) with linear regression as explain in the Method section.

The result of GSA is shown in Fig. 4. As can be easily seen, four parameters are sensitive: λ_1, λ_3, K, and m_2. Parameters λ_1 and λ_3 represents the speed of growth of the primary tumor and the distant metastases. It is thus not surprising that these parameters affect MFS as with higher λ_1 and λ_3 probability of dissemination to distant places and growth of distant metastases increase. Also, parameter m_2 is not surprisingly sensitive as it is the dissemination rate to distant metastases from the primary tumor.

What is perhaps less expected, parameter K which is carrying capacity is sensitive to MFS. It means, that reducing a constrain on the maximal tumor size could perhaps delay the emergence of distant metastases. From the clinical point of view, it means that reducing angiogenesis through angio-therapy such as VEGF inhibitor (for example, bevacizumab) could delay the emergence of metastases.

Interestingly, parameter u, which is a mutation rate responsible for transferring cancer cells into ones with metastatic potential, has a minor effect on MFS. It means, that EMT is not the dominant process in distant metastases formation in lung cancer.

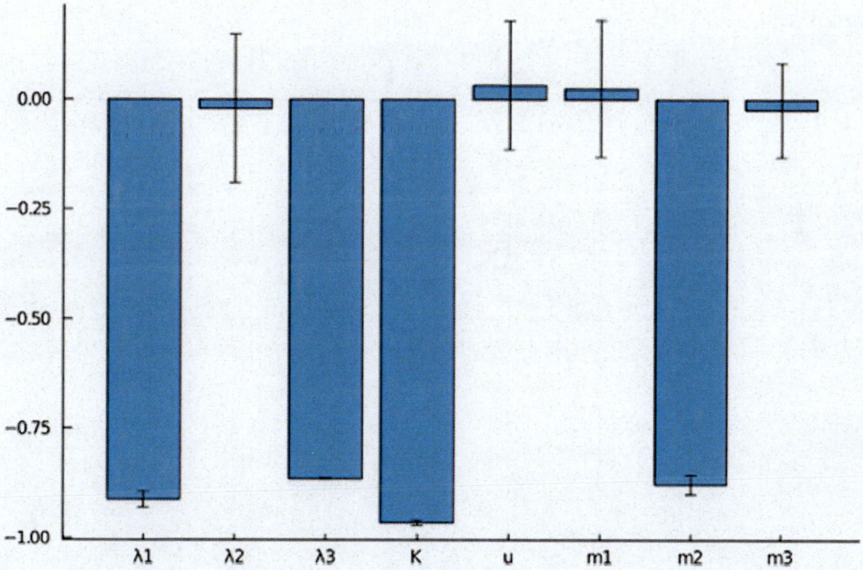

Fig. 4 Global sensitivity analysis in response to metastasis-free survival

4.4 The Probability of Metastases as a Function of Tumor Size at the Diagnosis Is Also Dependent on Carrying Capacity

In the last step, to further confirm that carrying capacity is indeed a key parameter that affects metastasis potential, we computed the probability of metastases as a function of tumor burden at diagnosis as follows. From a given tumor size which reflects the size of the primary tumor at diagnosis, 10,000 simulations are performed until metastasis is detected or for 2 years. Next, the probability of metastases is calculated as a sum of simulations when distant metastases are detected divided by all simulations:

$$probability = \left\{ \frac{1\ \mathit{if}\ D(t)\ \geq\ 10^9}{0\ \mathit{if}\ D(t)\ <\ 10^9} \right. : \mathrm{n} \tag{5}$$

The calculations were repeated for 100 different values of tumor size at the diagnosis. The results were then plotted with tumor size on the x-axis and probability on the y-axis.

To check how each parameter affects the probability of distant metastases, we decreased each parameter by 10% and plotted the probability of distant metastases as a function of tumor size at the diagnosis. The result of this analysis is shown in Fig. 5.

Only two parameters affect the shape of probability as a function of tumor burden at the diagnosis: λ_3 and K. It is not surprising that the growth rate of cancer cells in distant metastasis compartments is a key parameter. And indeed for λ_3 decreased by 10%, we can see no metastases within two years from diagnosis for a wide range of tumor size at diagnosis.

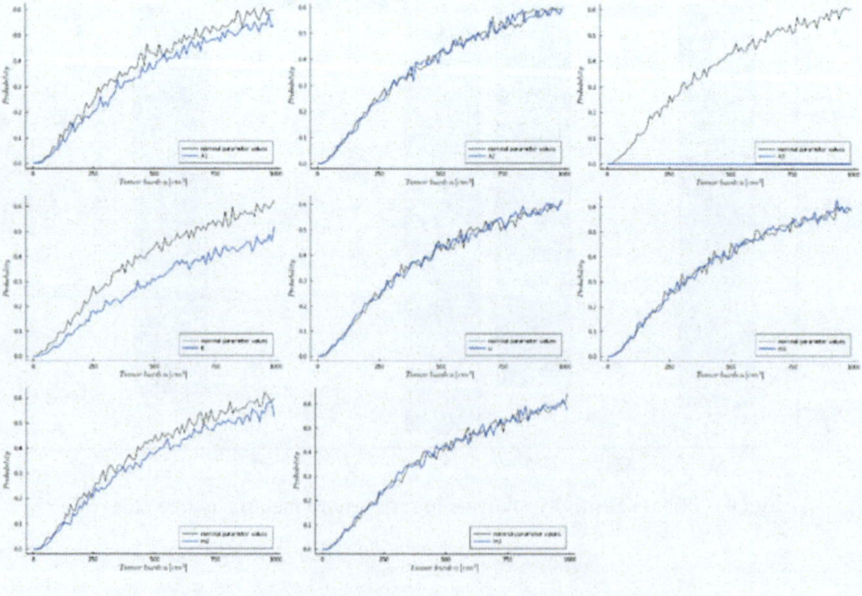

Fig. 5 Probability of metastasis as a function of tumor burden at the diagnosis for nominal parameters and decreased parameters by 10%

Interestingly, the probability of metastases as a function of tumor size at diagnosis depends on carrying capacity. Again, the model predicts that angio-therapy could delay the emergence of metastases.

5 Discussion

Lung cancer is one of the most metastatic cancers, and thus the most lethal ones. In contrast to other cancer types, patients with lung cancer die usually because of metastases rather than the primary tumor. Thus, it is important to develop novel strategies to overcome or at least delay the emergence of metastases. Here, we tackle the problem using mathematical oncology methods by using mathematical modeling.

We developed a stochastic model of primary lung cancer growth together with metastasis dissemination, and growth of distant and lymph nodes metastases. The model considers two ways of metastasis dissemination through lymphatic vessels and blood vessels. We constructed the model in form of stochastic differential equations (SDE) which allows for efficient simulation of a stochastic model. The model is implemented in Julia language using DifferentialEquations.jl library which is very efficient in solving SDE.

Using the developed model, we applied global sensitivity analysis to investigate which model parameters affect metastasis-free survival. We found four parameters that affect MFS: the growth rate of the primary tumor, the growth rate of distant metastases, dissemination rate from the primary tumor to distant metastases, and carrying capacity. From all four parameters, we can steer two of the growth rate of the primary tumor and carrying capacity. The first parameter we could steer using chemotherapy which is a common treatment modality in lung cancer. The second parameter could be steered using angio-therapy such as bevacizumab.

We suggest that angio-therapy could delay the emergence of metastases in lung cancer. This opens the possibility for the treatment of early-stage lung cancer patients with angio-therapy. For instance, Ilhan-Mutlu et al. performed *in vivo* research to check the effect of bevacizumab on the emergence of brain metastases [14]. They discovered that bevacizumab significantly reduces the probability of brain metastases. However, to our best knowledge, no clinical study was performed to check those findings. Here, using a mathematical oncology approach, we further strengthen the hypothesis that angio-therapy could delay metastasis dissemination to distant organs.

6 Future Directions

Our results are preliminary ones to further application of mathematical oncology approach to tackling the emergence of metastases in lung cancer as well as finding novel treatment methods to overcome metastases.

Presented herein the mathematical model is a simplified model of metastases dissemination and includes only eight parameters. In the reality, metastases creation is a complex process with five distinct steps. Also, in metastases are involved not only cancer cells but also immune cells, fibroblasts, and other types of cells which are present in the tumor microenvironment. Thus, we plan to expand the model in such a way that it includes all five steps of tumor dissemination as well as a different type of cells in the tumor microenvironment.

The developed model could be also modified to include angiogenesis by non-constant carrying capacity in a similar manner as the Hahnfeldt *et al.* model [15]. This could increase the reality of the growth of metastases. However, from another point of view, it will make the model more complex as at least two additional parameters need to be included. Yet another question is how the growth dynamics of metastasis could be described. We have decided to use Gompertz-type dynamics but a classical exponential growth dynamics or more sophisticated superlinear growth model [15] could be considered as alternative approaches.

Next, we also plan to integrate the model with clinical and PET/CT scans data. Clinical data will be applied to fit the model to the Kaplan–Meier plot where on the x-axis is metastasis-free survival. PET/CT scans, however, will be applied to find radiomic biomarkers that could predict the time to metastasis.

Acknowledgements. We thank Professor Rafał Suwiński from the National Cancer Institute in Poland, Gliwice Branch for fruitful discussions and assistance in the clinical part of the project. We would like to acknowledge the financial support of the National Science Center, Poland - grant number 2020/37/B/ST6/01959 (AS, EK) and Foundation for Polish Science (FNP) - scholarship START (EK).

References

1. Sung, H., Ferlay, J., Siegel, R.L., et al.: Global cancer statistics 2020: GLOBOCAN estimates of incidence and mortality worldwide for 36 cancers in 185 countries. CA Cancer J. Clin. 21660 (2021). https://doi.org/10.3322/caac.21660
2. Inamura, K.: Lung cancer: understanding its molecular pathology and the 2015 WHO classification. Front. Oncol. **7**, 193 (2017). https://doi.org/10.3389/fonc.2017.00193
3. Molina, J.R., Yang, P., Cassivi, S.D., et al.: Non-small cell lung cancer: epidemiology, risk factors, treatment, and survivorship. In: Mayo Clinic Proceedings. Elsevier Ltd., pp. 584–594 (2008)
4. Zhu, T, Bao, X, Chen, M., et al.: Mechanisms and future of non-small cell lung cancer metastasis. Front. Oncol. **10**, 2441 (2020). https://doi.org/10.3389/fonc.2020.585284
5. Fares, J., Fares, M.Y., Khachfe, H.H., et al.: Molecular principles of metastasis: a hallmark of cancer revisited. Signal. Transduct. Target. Ther. **5**, 1–17 (2020)
6. Popper, H.H.: Progression and metastasis of lung cancer. Cancer Metast. Rev. **35**, 75–91 (2016). https://doi.org/10.1007/s10555-016-9618-0
7. Fenske, D.C., Price, G.L., Hess, L.M., et al.: Systematic review of brain metastases in patients with non–small-cell lung cancer in the United States, European Union, and Japan. Clin. Lung Cancer **18**, 607–614 (2017)
8. Cho, Y.J., Cho, Y.M., Kim, S.H., et al.: Clinical analysis of patients with skeletal metastasis of lung cancer. BMC Cancer **19**, 303 (2019). https://doi.org/10.1186/s12885-019-5534-3
9. Raz, D.J., Lanuti, M., Gaissert, H.C., et al.: Outcomes of patients with isolated adrenal metastasis from non-small cell lung Carcinoma. Ann. Thorac. Surg. **92**, 1788–1793 (2011). https://doi.org/10.1016/j.athoracsur.2011.05.116
10. Fan, Y., Xu, X., Xie, C.: EGFR-TKI therapy for patients with brain metastases from non-small-cell lung cancer: a pooled analysis of published data. Onco. Targets Ther. **7**, 2075–2084 (2014). https://doi.org/10.2147/OTT.S67586

11. Zappa, C., Mousa, S.A.: Non-small cell lung cancer: current treatment and future advances. Transl. Lung Cancer Res. **5**, 288–300 (2016). https://doi.org/10.21037/tlcr.2016.06.07
12. Zill, O.A., Banks, K.C., Fairclough, S.R., et al.: The landscape of actionable genomic alterations in cell-free circulating tumor DNA from 21,807 advanced cancer patients. Clin. Cancer Res. **24**, 3528–3538 (2018). https://doi.org/10.1158/1078-0432.CCR-17-3837
13. Bilous, M., Serdjebi, C., Boyer, A., et al.: Quantitative mathematical modeling of clinical brain metastasis dynamics in non-small cell lung cancer. Sci. Rep. **9**, 13018 (2019). https://doi.org/10.1038/s41598-019-49407-3
14. Ilhan-Mutlu, A., Osswald, M., Liao, Y., et al.: Bevacizumab prevents brain metastases formation in lung adenocarcinoma. Mol. Cancer Ther. **15**, 702–710 (2016). https://doi.org/10.1158/1535-7163.MCT-15-0582
15. Hahnfeldt, P., Panigrahy, D., Folkman, J., Hlatky, L.: Tumor development under angiogenic signaling. Cancer Res. **59**, 4770–4775 (1999)

Check for updates

The Influence of Hip Conditioning Program with Rotational Movements on Thermal Response of Lower Limbs

Jolanta G. Zuzda[1], Jakub Kacpura[2], Jakub Dziura[2(✉)], Manuel Sillero Quintana[3], and Robert Latosiewicz[4]

[1] Tourism and Recreation Department, Bialystok University of Technology, 16-001 Bialystok, Poland
fitness2004@wp.pl
[2] Rehabilitation Center "Sprawność", 05-220 Zielonka, Poland
[3] Departmento de Deportes, Universidad Politecnica de Madrid, Calle Martín Fierro, 7, 28040 Madrid, Spain
manuel.sillero@upm.es
[4] Department of Rehabilitation and Physiotherapy, Medical University of Lublin, ul. Magnoliowa 2, 20-093 Lublin, Poland
rlatos@umlub.pl

Abstract. Purpose: The purpose of this study was the evaluation of differences in skin temperature on the body surface and of the impact of level of physical activity, characteristics of leg muscles and flexibility on the skin temperature during Hip Conditioning Program with Rotational Movements (HCP).

Methods: The study included 35 healthy volunteers of the Bialystok University of Technology. A thermographic camera and dedicated software were used for the acquisition and processing of thermographic images of body muscles before and after the HCP training regime.

Results: Hip Conditioning Program with Rotational Movements results in alignment of thermoregulatory processes in groups with lower fitness level and lower lean muscle mass in comparison to groups better developed in this matter. Individuals which has less developed muscles shows higher differences in skin temperature than groups more fit and muscle-developed.

Conclusion: Hip Conditioning program with Rotational Movements proved to be one of forms of physical activity that are suited to track and study the change of skin temperature measured by thermographic camera in subjects with different levels of lean muscle mass.

The usage of infrared thermography proved to be potent investigation tool in assessing relationship between thermoregulatory processes and forms of physical activity with taking into account the individual traits and characteristics of participants.

Keywords: Thermography · Hip Conditioning Program · Skin temperature · Muscles · Flexibility · Physical activity

D. G. Pijanowska et al. (eds.), *Biocybernetics and Biomedical Engineering – Current Trends and Challenges*, Lecture Notes in Networks and Systems 293,
https://doi.org/10.1007/978-3-030-83704-4_8

1 Introduction

Healthy individuals are crucial for modern society, as without people who can work until their retiring age in good mental and physical health, there won't be enough people to ensure stable and dynamic progress in science, technology and global society. It is a common knowledge that many forms of sports, fitness and physiotherapy in general could elevate one's health In brief, we could mention three forms of physical activities (PA) that improve our well-being like Cardiorespiratory Fitness and Weight Control Exercise, Muscular Strength and Muscular Endurance and Flexibility Exercise. Flexibility exercises multiple benefits including: improving joint ROM and function and enhancing muscular performance. The American college of Sports Medicine recommendations for the quantity and quality of training for developing and maintaining cardiorespiratory fitness, body composition, muscular strength and endurance, and flexibility in the healthy adult are described by Pollock et al. [1]. Keeping people, especially children and adolescents, in good overall shape and health is crucial for avoiding increased development of obesity in global population as it's becoming more of a problem in both developed and developing countries. Moreover, obesity in adolescence leads to exponentially increased risk of cardiovascular diseases and diabetes mellitus in adulthood [2, 3].

Hip Conditioning Program with Rotational Movement (HCP) consists of simple movements and low/medium intensity. In this kind of movements, the recruitment of muscle groups occurs that usually are not stimulated in the linear movements. Moreover, rotational movements can improve the structure and functions of the ligaments, tendons and joints, increasing flexibility. Despite of the fact that rotation exercises are commonly simple exercises involving rotational movements of large majority of body segments and allowing the training of neuro-muscular systems that are usually neglected in most common approach in training. These exercises improve muscle efficiency, coordination and posture, benefiting the client's health status [4].

The physiological effects of HCP are reasonably well established in terms of cardiovascular activity, muscular fitness, changes in body composition and bone health after mechanical load and musculoskeletal injuries. Still, the impact of level of physical activity, leg muscles and flexibility on the skin temperature during HCP have not yet been assessed and described.

Thermography is a method which can be used to register body thermal patterns in non-invasive way [5, 6]. Furthermore, infrared thermography (IRT) now transformed into a potent investigative tool with many possible applications starting from historical purposes such as mechanical, electrical or military to more modern approaches as in architecture and medical technology or even astronomy. It is safe method, as it detects normal, thermal radiation and it could be used in all living creatures, to monitor physiological and anatomical structures based on heat emitted by skin. The human body is described as homeothermic–it means it has constant inner temperature regardless of external factors like humidity or outside temperature (i.e. 36.8 ± 0.5 °C) which is optimal for the best organism functioning [7]. Thermoregulation is a complex system consisting not only of body tissues like skin, but also blood vessels and sympathetic nervous system which manages this processes throughout cholinergic and adrenergic system [8]. The image produced with thermal imaging cameras is called thermogram and utilizes radiation in the infrared range of the electromagnetic spectrum which it detects. IRT can provide a researcher with real-time visual image with accurate measurement of surface temperature over the wide body area.

Based on its non-invasiveness, IRT has been defined as the scientific analysis of data from non-contact thermal imaging devices [6–9]. There are number of articles in this matter. Thermography could be used to assess cumulative training loads and check how body can cope with muscle fatigue, thus monitor exercise recovery [9]. The muscle temperature could also be a forecast for trainers to measure possible outcomes from training regimen, as increase in temperature of muscle, could decrease efficiency of working muscle, thus it could be helpful to plan whole muscle strengthening process [10]. As we can see usage of IRT could be crucial to objectively measure and help scientists and researchers to enlarge efficacy of their approaches regarding body function, performance and preventive means, especially in the field of sports medicine and rehabilitation where there is a great need for effective methods and approaches to maximize individuals' health and performance.

2 Purpose

The aim of this study is to analyze the evolution of changes in skin temperature (Tsk) on selected body areas (Regions of Interest–ROI) of persons practicing Hip Conditioning Program with Rotational Movements (HCP) and of the impact of level of physical activity, leg muscles characteristics and flexibility on the thermoregulatory system.

3 Materials and Methods

3.1 Subjects

A total of thirty five employees from the Bialystok University of Technology (BUT) in Bialystok, Poland (age: 48.89 ± 11.53 [y]; height: 1.67 ± 0.08 [m]; weight: 69.89 ± 12.53 [kg]; Body Mass Index (BMI): 24.96 ± 3.75 [kg/m^2]; systolic blood pressure (SBP): 126.82 ± 11.23 [mm Hg]; diastolic blood pressure (DBP):75.89 ± 6.68 [mm Hg]) reporting that they would exercise at least two times per week volunteered for this study. Inclusion criterion was good general health status, which would allow to participate in the HCP. Before HCP, all subjects were informed about the training protocol and gave their informed consent regarding study procedures. The Physical Activity Readiness Questionnaire (PAR–Q+) was used to provide information relevant to the safety of beginning HCP and to identify known diseases and risk factors for cardiovascular diseases or orthopedic injury. This screening instrument was used to identify potentially dangerous health conditions that contraindicate physical activity by a medical doctor prior to beginning the exercises. Also, candidates should have no physical limitations or diseases, and declared no consumption of drugs, alcohol or tobacco.

3.2 Hip Conditioning Program with Rotational Movements and Establishment of Working Load Intensity parameters

HCP training was conducted by the author of this program (JGZ). Sessions were held at the air-conditioned BUT Sports Hall with a constant temperature of 22 ± 0.7 °C,

$50 \pm 5\%$ humidity and under artificial light of constant intensity. Training unit consisted of a warm-up routine, the main part routine, and stretching and breathing exercise. The intensity of HCP exercises was at about 60–70% of maximal heart rate (MHR), The subjective intensity of the HCP was measured using the Borg scale of perceived exertion (from 6 to 20) [12]. In the morning of the day of training, subjects recorded their individual resting heart rate (HR), and this value was used to calculate their MHR with use of the Whaley formula [13].

3.3 Procedure

The study was conducted at the Bialystok University of Technology (BUT) in Bialystok, Poland and was approved by the local Bioethics Committee (KE-0264/198/2016). During the first visit all the participants completed written informed content the PAR-Q+. All participants were tested for each of the following physical fitness components: height, body mass, body mass index, body composition. Each subject's body height was measured by using professional scale and stadiometer WPT 100/200 OW (RADWAG Wagi Elektroniczne, Poland). During the second visit all the participants completed IPAQ–SF protocol, Body Composition Protocol, Flexibility test and Thermographic protocol pre– and post–session of HCP.

3.4 Body Composition Protocol

Participants' body composition was measured by bioelectrical impedance assay (BIA) using a Jawon IOI 353 Segmental Body Composition Monitor with Jawon Medical software. Body Composition Analyzer JAWON MEDICAL IOI-353 (Accuniq, South Korea) is EC0197 certified and meets the requirements of MDD 93/42EEC in the field of medical devices [15]. To perform an appropriate analysis all of the subjects were required to keep following rules before the BIA: no food ingestion for at least 4 h, minimal intake of 2 L of water the day before, no physical activity for at least 8 h, no coffee or alcoholic beverage consumption during at least 12 h, and no diuretic use for at least 24 h [14]. Participants were also asked to empty their bladder immediately before the BIA test. The device analyzed body composition based on the differences of the ability to conduct the electrical current by body tissues–different resistance. Body mass and body composition measurements taken with the use of the Jawon IOI 353 were as follows: weight (kg), height (cm), Lean Body Mass (LBM, kg), Soft Lean Mass (SLM, kg), Total Body Water (TBW, kg), Body Fat (kg), Percent Body Fat (%), Body Mass Index (BMI kg/m^{-2}), Blood Pressure: Systolic and Diastolic (mm Hg) and segmental assessment: Mass of Body Fat (MBF, kg) and Soft Lean Mass (SLM, kg) for (Lt Arm), (Rt Arm), (Trunk), (Lt Leg) and (Rt Leg).

3.5 Flexibility Test Protocol [16]

The Flexitest is a simple and effective method used to measure and assess flexibility for a wide variety of professional purposes [16]. Soares de Araujo stressed that Flexitest is a simple, easy and that simultaneously measures, evaluates for a wide variety of professional purposes the flexibility of 20 different joints individual movements [16]. In a simplistic analysis, flexibility scores were evaluated according to the following scale: 0–very poor; 1–poor; 2–average; 3–good; 4–very good, for each of the movements and joints. Soares de Araujo stressed that it is valid and appropriate to add the results for each of 20 movements and get overall flexibility index called Flexindex [16]. Flexindex equals the sum of 20 individual mobility scores. To perform a proper analysis, all subjects were required to follow the following principles before the Flexitest: not exercising within the hour before the test; being comfortable in their clothing and removal of jewelry. They were informed before the measurements about position they should take for the measurements and also explained that some level of discomfort could be expected when doing the test, especially in the arms and wrists.

3.6 Level of Physical Activity Protocol

To assess the level of PA participants completed the short version of the International Physical Activity Questionnaire (IPAQ–SF) recalling their PA during the seven-day period [17]. In this questionnaire there are determined separately the number of days and the time devoted to PA. The measure of energy consumption during PA is the so–called MET (Metabolic Equivalent). PA expressed in MET-min/week is calculated by multiplying the number of days a certain effort was performed by the value of MET for this effort and by the average number of minutes the effort was performed a day [17, 18]

3.7 Thermographic Protocol

A thermographic camera (CEDIP Titanium 560, Sofradir EC Inc., USA) with a real integrated resolution of 640×512 pixels and sensors measuring temperatures ranging from -20 °C to $+3000$ °C was used for thermal image acquisition following the TISEM recommendations. The camera was sensitive enough to detect differences in temperature lower than 15 mK and had accuracy of ± 1 °C of absolute temperature [19]. Images were taken with an emissivity of 0.98. A specific software for thermographic image processing (AltairTM 5.80) was used to select manually the ROI. For verification purposes, a remote infrared thermometer CEM DT–8819 (Poland) was also used.

Thermographic images were taken before and immediately after exercise in an air-conditioned nearby room with a constant temperature of 21.8 ± 0.09°C and humidity of $50 \pm 5\%$ after an acclimation period to the room conditions, that lasted at least 10 min and no longer than 20. Measurements were taken at 10 m from the subject. Two series of thermograms were taken; the first series of thermograms were taken before starting the HCP exercises and the second series was taken after HCP exercises (Fig. 1). For evaluation purposes six ROI were defined in the anterior of the body: Thigh left (AL Thigh), Thigh right (AR Thigh), Knee Left (AL Knee), Knee Right (AR Knee), Calf left (AL Calf), Calf Right (AR Calf). Six ROI were defined in the posterior of body: Thigh left (PL Thigh), Thigh right (PR Thigh), Knee Right (PR Knee), Calf left (PL Calf), Calf Right (PR Calf). The ROIs were chosen and grouped corresponding to anatomical parts of the human body. For calves, knees and thighs, measurements were taken symmetrically on the left and right side of the body (Table 1).

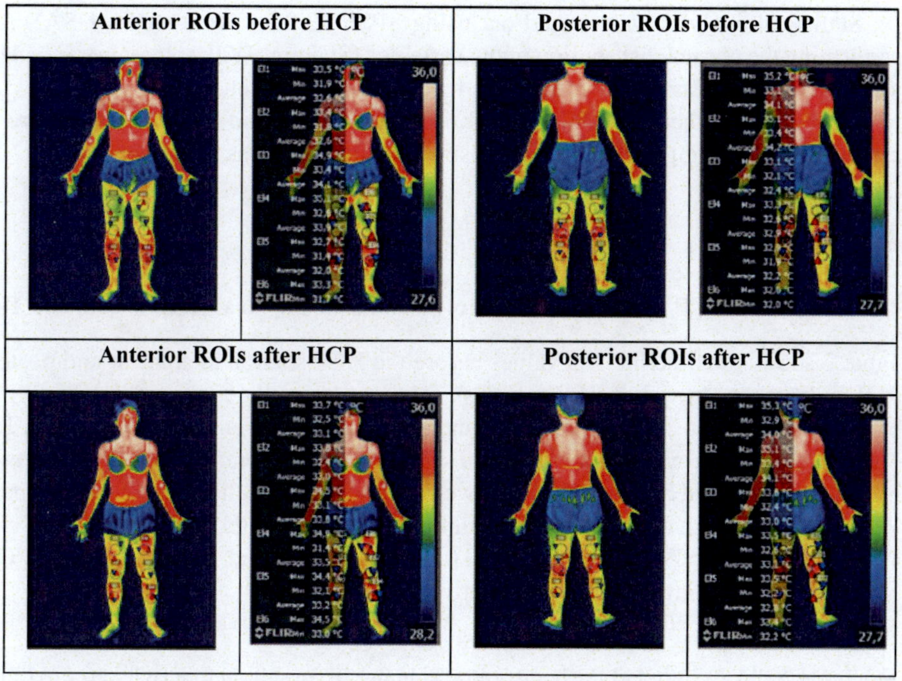

Fig. 1 The placement of ROI at the anterior and posterior part of the body

Table 1 Region of Interest (ROI) and corresponding muscles or groups of muscles

Body site	ROI	Muscles, ligaments, tendons their parts or groups	Integrated ROI*
Anterior (Ant.)	Ant. Thigh	m.quadriceps femoris, m.rectus femoris, m.sartorius	AL and AR Thigh
	Ant. Knee	patellar tendon, tendon of quadriceps femoris	AL and AR Knee
	Ant. Calf	m. tibialis anterior, m. extensor digitorum longus, m.extensor hallucis longus	AL and AR Calf
Posterior (Post.)	Post. Thigh	m. biceps femoris, m. semitendinosus, m. semitendinosus	PL and PR Thigh
	Post. Knee	m.gastrocnemius, tendon of m. biceps femoris, tendon of m. semitendinosus	PL and PR Knee
	Post. Calf	m.soleus, m.gastrocnemius, Achilles tendon	PL and PR Calf

* – abbreviations are explained in section “Materials and methods”

The distribution of evaluated muscle groups over body figure and selected ROI of measurement are shown in Fig. 1.

Statistical analysis was carried out using SPSS 23.0 software (IBM, USA). After confirming the normal distribution of the variables (Shapiro–Wilk test), it was decided to use Student t test of non-related samples for evaluating differences between participants with lower and higher levels of MET, muscle of the legs and flexibility in before and after the HCP training session. The statistical significance was set at $\alpha = 0.05$.

4 Results

4.1 Effects of HCP on Tsk and Impact of Level of Physical Activity, on the Tsk

Table 2 shows the baseline Tsk values before the HCP session in anterior and posterior ROI and impact of level of PA on the Tsk. A summary of the averaged Tsk and SD for each anterior and posterior ROI before and after HCP is presented in Table 2 and Table 3. From these data we concluded that Tsk significant difference ($p < 0.05$) between groups with low and high MET results were observed at baseline (Table 2) only in Tsk ROI Ant. Left Thigh and Post. Right Thigh. Average Tsk in those ROI was 29.07 °C ± 1.94 °C vs. 30.08 °C ± 0.92 °C ($p \leq 0.05$), respectively and 30.08 °C ± 1.73 °C vs. 31.04 °C ± 0.82 °C ($p = 0.036$), respectively.

Table 2 Tsk (°C) ROI of the body of low vs high MET results before HCP (*p < 0.05)

Before exercise	Low (N = 15)		High (N = 20)		Difference	t (33)	p
	Average mean	SD	Average mean	SD			
AR Knee	30,04	1,46	29,74	0,86	0,30	0,755	0,456
AL Knee	29,97	1,59	29,85	0,90	0,12	0,288	0,776
AR Leg	31,93	1,15	32,11	0,58	−0,18	−0,619	0,540
AL Leg	31,71	1,44	32,08	0,64	−0,37	−1,023	0,314
AR Thigh	29,08	1,96	30,06	0,92	−0,98	−1,973	0,057
AL Thigh	29,07	1,94	30,08	0,92	−1,00	−2,039	**0,050**
PR Knee	31,48	1,03	31,88	0,67	−0,40	−1,377	0,178
PL Knee	31,29	1,12	31,85	0,75	−0,56	−1,768	0,086
PR Leg	30,75	2,06	30,89	0,89	−0,14	−0,272	0,787
PL Leg	30,35	2,12	30,66	0,69	−0,31	−0,613	0,544
PR Thigh	30,08	1,73	31,04	0,82	−0,96	−2,184	**0,036**
PL Thigh	30,15	1,77	30,81	0,79	−0,66	−1,492	0,145

Table 3 shows the baseline Tsk values after the HCP session in anterior and posterior ROI and impact of level of physical activity, on the Tsk. After the HCP training there was no significant difference ($p > 0.05$) between groups for all measured areas in anterior and posterior ROI.

Table 3 Tsk (°C) ROI of the body of low vs. high MET results after HCP *p < 0.05

After exercise	Low (N = 10)		High (N = 19)		Dif	$t_{(33)}$	p
	Average mean	SD	Average mean	SD			
AR Knee	30,67	0,95	30,29	0,84	0,38	1,098	0,282
AL Knee	30,65	0,80	30,30	0,85	0,34	1,054	0,301
AR Leg	32,02	0,58	31,93	0,64	0,09	0,354	0,726
AL Leg	31,72	0,71	31,84	0,66	−0,12	−0,449	0,657
AR Thigh	30,58	0,86	31,02	0,64	−0,44	−1,565	0,129
AL Thigh	30,62	0,92	30,96	0,69	−0,34	−1,123	0,271
PR Knee	31,99	0,70	31,93	0,86	0,06	0,180	0,859
PL Knee	31,77	0,46	31,86	0,82	−0,09	−0,327	0,746
PR Leg	31,02	1,44	30,85	1,00	0,17	0,366	0,718
PL Leg	30,44	1,07	30,69	0,83	−0,25	−0,686	0,499
PR Thigh	30,88	1,00	31,17	0,79	−0,29	−0,849	0,403
PL Thigh	30,98	1,22	31,02	0,98	−0,03	−0,080	0,937

4.2 Effects of HCP on Tsk and Impact of Soft Lean Muscle Mass of Muscles in Lower Limbs

Table 4 and 5 shows the baseline Tsk values before the HCP session and after the training program and impact of level of soft lean muscle mass, on the Tsk. Skin temperature level show no significant difference ($p > 0.05$) between groups with Low and High SLM results that were observed at baseline (Table 4).

Table 5 shows the baseline Tsk values after the HCP session in anterior and posterior ROI. And impact of level of muscles of legs, on the Tsk. After HCP significant difference ($p < 0.05$) between groups with Low and High SLM results were observed (Table 5) only in Tsk ROI Ant. Right Thigh and Post Left Thigh. Average Tsk in this ROI was 31.12 °C ± 0.69 °C ($p = 0.13$) vs. 30.44 °C ± 0.64 °C ($p = 0.017$), respectively and 31.11 °C ± 0.74 °C vs. 30.42 °C ± 0.68 °C ($p = 0.036$), respectively.

4.3 Effects of HCP on Tsk and Impact of Level of Flexibility on the Tsk

Table 6 and 7 shows the baseline Tsk values before the HCP session and after the training program and impact of level of flexibility on the Tsk. From these data Significant difference ($p < 0.05$).between groups with Low and High Flexibility results were observed at baseline (Table 6) in Tsk ROI Ant. Right Calves in this ROI average Tsk was 31.67 °C ± 0.98 °C vs. 32.37 °C ± 0.57 °C ($p = 0.013$), respectively and Ant. Left Calves in this ROI average Tsk was 31.48 °C ± 1.19 °C vs. 32.33 °C ± 0.72 °C, ($p + 0.015$) respectively. In the Post. Left Knees average Tsk was 31.28 °C ± 1.11 °C vs. 31.92 °C ± 0.67 °C ($p = 0.044$), respectively.

Table 4 Tsk (°C) ROI of the body of low vs. high SLM level muscles of legs results before HCP *p < 0.05

Before exercise	Low (N = 23)		High (N = 12)		Dif	T (33)	p
	Average mean	SD	Average mean	SD			
AR Knee	29,77	1,04	30,05	1,36	−0,28	−0,675	0,504
AL Knee	29,80	1,06	30,10	1,53	−0,30	−0,674	0,505
AR Leg	31,95	0,72	32,18	1,11	−0,23	−0,751	0,458
AL Leg	31,97	0,76	31,83	1,51	0,14	0,360	0,721
AR Thigh	29,72	1,25	29,49	1,98	0,23	0,424	0,674
AL Thigh	29,74	1,22	29,48	2,00	0,26	0,474	0,638
PR Knee	31,63	0,85	31,86	0,88	−0,23	−0,742	0,463
PL Knee	31,61	0,98	31,61	0,95	0,01	0,022	0,982
PR Leg	30,60	0,94	31,26	2,17	−0,66	−1,259	0,217
PL Leg	30,54	1,03	30,50	2,12	0,04	0,082	0,935
PR Thigh	30,61	1,09	30,66	1,82	−0,05	−0,108	0,915
PL Thigh	30,58	1,04	30,44	1,79	0,14	0,292	0,772

Table 5 Tsk (°C) ROI of the body of low vs high muscles of legs results after HCP *p < 0.05

After exercise	Low (N = 18)		High (N = 11)		Dif	t (33)	p
	Average	SD	Average	SD			
AR Knee	30,51	0,95	30,29	0,78	0,22	0,642	0,526
AL Knee	30,54	0,86	30,23	0,81	0,31	0,971	0,340
AR Leg	31,87	0,57	32,12	0,68	−0,25	−1,054	0,301
AL Leg	31,87	0,65	31,68	0,71	0,19	0,733	0,470
AR Thigh	31,12	0,69	30,44	0,64	0,68	2,663	**0,013**
AL Thigh	31,11	0,74	30,42	0,68	0,69	2,540	**0,017**
PR Knee	31,95	0,78	31,95	0,86	−0,01	−0,026	0,979
PL Knee	31,91	0,76	31,71	0,61	0,20	0,735	0,469
PR Leg	30,73	0,90	31,21	1,47	−0,49	−1,110	0,277
PL Leg	30,67	0,91	30,48	0,94	0,19	0,534	0,598
PR Thigh	31,19	0,84	30,88	0,90	0,31	0,936	0,358
PL Thigh	31,11	1,00	30,84	1,15	0,27	0,675	0,505

Table 6 Tsk (°C) ROI of the body of low vs. high flexibility of legs results before HCP *p < 0.05

Before exercise	Low (N = 23)		High (N = 12)		Dif	t(33)	p
	Average	SD	Average	SD			
AR Knee	29,62	1,20	30,10	1,07	−0,48	−1,247	0,221
AL Knee	29,53	1,25	30,26	1,13	−0,73	−1,819	0,078
AR Leg	31,67	0,98	32,37	0,57	−0,70	−2,615	**0,013**
AL Leg	31,48	1,19	32,33	0,72	−0,85	−2,572	**0,015**
AR Thigh	29,34	1,60	29,93	1,42	−0,59	−1,150	0,258
AL Thigh	29,46	1,71	29,83	1,32	−0,37	−0,720	0,476
PR Knee	31,48	1,03	31,92	0,60	−0,43	−1,532	0,135
PL Knee	31,28	1,11	31,92	0,67	−0,64	−2,097	**0,044**
PR Leg	30,80	1,84	30,85	1,09	−0,05	−0,100	0,921
PL Leg	30,31	1,80	30,73	1,07	−0,43	−0,858	0,397
PR Thigh	30,36	1,53	30,88	1,15	−0,52	−1,133	0,265
PL Thigh	30,37	1,58	30,68	1,04	−0,31	−0,679	0,502

Table 7 Tsk (°C) ROI of the body of low vs. high flexibility of legs results after HCP *p < 0.05

After exercise	Low (N = 18)		High (N = 11)		Dif	t (33)	p
	Average	SD	Average	SD			
AR Knee	30,19	1,00	30,64	0,71	−0,45	−1,403	0,172
AL Knee	30,21	0,93	30,62	0,72	−0,41	−1,319	0,198
AR Leg	31,79	0,62	32,12	0,58	−0,33	−1,452	0,158
AL Leg	31,48	0,61	32,09	0,59	−0,60	−2,712	**0,012**
AR Thigh	30,75	0,95	30,97	0,48	−0,22	−0,808	0,426
AL Thigh	30,75	0,95	30,94	0,60	−0,19	−0,663	0,513
PR Knee	31,75	0,97	32,13	0,56	−0,39	−1,322	0,197
PL Knee	31,59	0,77	32,06	0,57	−0,47	−1,893	0,069
PR Leg	30,92	1,44	30,90	0,85	0,02	0,055	0,957
PL Leg	30,38	0,97	30,81	0,84	−0,42	−1,265	0,217
PR Thigh	30,85	0,91	31,28	0,79	−0,43	−1,362	0,184
PL Thigh	30,87	1,20	31,13	0,91	−0,26	−0,658	0,516

Table 7 shows the baseline Tsk values after the HCP session and impact of level of flexibility on the Tsk after HCP significant difference ($p < 0.05$) between groups with Low and High Flexibility results were observed only in Tsk ROI Ant. Left Calf. Average Tsk in this ROI was 31.45 °C ± 0.61 °C vs. 32.09 °C ± 0.59 °C ($p = 0.012$), respectively.

5 Discussion

Many forms of physical activity (PA) are known to be a fundamental in improving people's health–physical and mental and overall fitness of one's body. Its great impact and relative low cost for an individual could raise overall general health in the population [4]. Hip Conditioning Program with Rotational Movements (HCP) thanks to novel approach could be recommend as a supplementary PA to all individuals to strengthen muscles, improve neuromuscular systems and increase flexibility.

Thanks to increased metabolic heat produced during PA triggering person's thermoregulatory mechanisms of excessive heat elimination which is managed by two processes: skin vasoconstriction, caused by the blood flow demand to active muscles and also by the skin vasodilation needed to increase blood flow and heat conduction towards the skin [20]. Evaporation is the most effective mechanism for eliminating excess heat, thus sweating is the essential in that matter [10].

Studies of thermal responses to different forms of PA prove that depending on form of PA we could observe increases or decreases in Tsk directly afterwards. HCP resulted in no significant difference in groups with low versus high SLM regarding Tsk difference before exercise. After completing the HCP regime observations have shown temperature drop in the anterior ROI of thigh region: for Ant. Right Thigh (31.12 °C ± 0.69 °C ($p = 0.13$) vs. 30.44 °C ± 0.64 °C ($p = 0.017$) and for Ant. Left Thigh 31.11 °C ± 0.74 °C vs. 30.42 °C ± 0.68 °C ($p = 0.036$).

Groups' distribution depending on level of physical activity presents differences only before HCP. Tsk significant difference ($p < 0.05$) between groups with Low and High MET results were observed at baseline only in ROI Ant. Left Thigh and Post. Right Thigh. Averages Tsk in this ROI were 29.07 °C ± 1.94 °C vs. 30.08 °C ± 0.92 °C ($p \leq 0.05$) and 30.08 °C ± 1.73 °C vs. 31.04 °C ± 0.82 °C ($p = 0.036$), respectively. After HCP there was no difference between before and after intervention by MET groups.

Tanda in 'The use of infrared thermography to detect the skin temperature response to physical activity' concludes that a well-trained individual would be capable of tolerating higher temperatures or decrease the rising metabolic heat amount, thanks to process of oxidation, compared to a less trained person [20]. This is in line with our findings related to Tsk differences in people with low and high SLM. It also should be noted that in the same article marked the fact that delay of skin dilation did not hindered oxygen supply within the blood supplying active muscles. Merla et. al in article [10] from 2010 and Chudecka and Lubkowska in articles from 2012 [21] reported that elevated physical efficiency results in better heat removal ipso facto increasing blood flow through the skin. Merla et. al also conclude that physiological explanation for that phenomenon is increase in arterial caliber, which improve vasodilatory mechanism and lowers vasomotor tone in body [10].

Kenney and Johnson [11] in their article stated that Tsk in groups consisted of untrained and trained participants over the certain muscle groups' decrease during exhausting exercises. Underlying mechanism of this is continuous vasoconstrictor response due to elevation of catecholamines and vasoconstricting hormones levels over exercise intensity raise. Zontak [22] stated that the type of performed exercise could change the dynamic of the Tsk change. His work showed that resistance training especially in initial phase of muscle work cause better vasoconstriction capability and produce better distribution of the endogenous heat during activity over the skin layers. Our study showed that exercise regimen can cause better adaptation to physical activity, thus better heat distribution.

There is still insufficient number of publications concerning role of thermographic imaging usage in evaluation of efficiency of different forms of physical activity and its impact on individual's health [23]. We think that thermography could be a great asset for researchers, health workers and can be used as a comparative and objective tool. There are many examples of using this method in animals and evaluating theirs body reaction to physical strain. Clark et. al used it to monitor Tsk changes of runners on the treadmill–they found that there is an immediate decrease in Tsk to reach steady value within 15 min after the start [24].

Regarding last grouping criterion which was individual's flexibility significant difference ($p < 0.05$) between groups with Low versus High Flexibility results before exercise regime were observed at baseline in Tsk ROI Ant. Right Calves [31.67 °C ± 0.98 °C vs. 32.37 °C ± 0.57 °C ($p = 0.013$)] and Ant. Left Calves [31.48 °C ± 1.19 °C vs. 32.33 °C ± 0.72 °C, ($p = 0.015$)]. In the Post. Left Knees Tsk average was 31.28 °C ± 1.11 °C vs. 31.92 °C ± 0.67 °C ($p = 0.044$), respectively. After the course, there were observed Tsk differences only in ROI Ant. Left Calf [31.45 °C ± 0.61 °C vs. 32.09 °C ± 0.59 °C ($p = 0.012$)].

In our opinion, the HCP caused a body adaptation related to the activation of parasympathetic nervous system and better performance and activity of certain regions related to hip and lower limbs. It showed also that level of physical fitness is not a factor which could impact results. Furthermore, the HCP could be a great help for people with lower physical fitness to increase the lower limbs performance.

Additionally, levels of SLM (low SLM vs. high SLM) and results of Tsk differences shows, that individuals which completed the HCP exercise regime could benefit from it, but the study demonstrates higher raise in Tsk in individuals with lower SLM than in people with higher SLM.

Flexibility is also a factor that can impact the results, as the data shows that after the HCP Tsk results between the low and high flexibility group are more homogeneous.

As in previous studies, our results support that phenomena associated with thermoregulatory system as removal of endogenous heat, state of homeostasis after exercises or even the level of locomotor load generated by exercise could be evaluated using thermography [10].

6 Conclusions

1. Hip Conditioning Program with Rotational Movements resulted in an alignment of Tsk of lower limbs ROI in low level physical fitness group and high level physical fitness group which could mean that thermoregulatory mechanism are more efficient for individuals that have high level of physical fitness.

2. Raise in Tsk after the HCP regime in anterior part of thighs is higher in people with low lean muscle mass than in people with high one, which could mean, that thermoregulatory mechanisms are more adapted in more developed individuals, in terms of lean muscle mass.
3. Usage of infrared thermography seems an interesting tool to assess relationship between thermoregulatory processes, level of physical activity and individuals features.
4. Thermography is safe, non-invasive method to investigate and evaluate efficacy of Hip Conditioning program with Rotational Movements.

References

1. Pollock, M.G., Glenn, A., Butcher, J.D., Després, J.P., Dishman, R.K., Franklin, B.A., Garber, C.E.: ACSM position stand: the recommended quantity and quality of exercise for developing and maintaining cardiorespiratory and muscular fitness, and flexibility in healthy adults. Med. Sci. Sports Exerc. **30**(6), 975–991 (1998)
2. Ho, H.C.H., Maddaloni, E., Buzzetti, R.: Risk factors and predictive biomarkers of early cardiovascular disease in obese youth. Diabetes Metab. Res. Rev. **35**, e3134 (2019). https://doi.org/10.1002/dmrr.3134
3. Lee, E.Y., Yoon, K.H.: Epidemic obesity in children and adolescents: risk factors and prevention. Front. Med. **12**, 658–666 (2018). https://doi.org/10.1007/s11684-018-0640-1
4. Zuzda, J.G., Latosiewicz, R.: Innovative program of recreational classes using rotational movements. Białystok University of Technology (2010)
5. Fernández-Cuevas, I., Marins, J.C.B., Arnaiz-Lastras, J., Carmona, P.M.G., Piñonosa Cano, S., García-Concepción, M.A., Sillero-Quintana, M.: Classification of factors influencing the use of infrared thermography in humans: a review. Infrared Phys. Technol. **71**, 28–55 (2015)
6. Hildebrandt, C., Zeilberger, K., Ring, E.F.J., Raschner, C.: The Application of Medical Infrared Thermography in Sports Medicine. InTechOpen (2012). https://www.intechopen.com/books/an-international-perspective-on-topics-in-sports-medicine-and-sports-injury/the-application-of-medical-infrared-thermography-in-sports-medicine. Accessed 2 Oct 2019
7. Hoover, K.C., Burlingame, S.E., Lautz, C.H.: Opportunities and challenges in concrete with thermal imaging. Concr. Int. **26**, 23–27 (2004)
8. Liu, Y., Mimura, K., Wang, L., Ikuda, K.: Physiological benefits of 24–style Tai–Chi exercise in middle–aged women. J. Physiol. Anthropol. Appl. Human Sci. **22**, 219–225 (2003)
9. Chudecka, M., Lubkowska, A.: Evaluation of the body surface temperature changes in the basketball players' after training. Acta Bio-Opt. Inform. Med. **4**, 271–274 (2011)
10. Merla, A., Mattei, P.A., Di Donato, L., Romani, G.: Thermal imaging of cutaneous temperature modifications in runners during graded exercise. Ann Biomed. Eng. **38**, 158–163 (2010)
11. Kenney, W.L., Johnson, J.M.: Control of skin blood flow during exercises. Med. Sci. Sports Exerc. **24**, 303–312 (1992)
12. Borg, G.A.: Psychophysical bases of perceived exertion. Med. Sci. Sports Exerc. **14**, 377–381 (1982). http://www.la.awf.poznan.pl/files/Biegi_wytrzym_jogging/Skala_Borga_dla_doros%C5%82ych.pdf. Accessed 2 Oct 2019
13. Whaly, M.A., Kaminsky, L.A., Dwyer, G.B., Getchell, L.H., Norton, J.: Predictors of over and under achievement of age predicted maximal heart rate. Med. Sci. Sports Exerc. **24**(10), 1173–1179 (1992)

14. Schiavo, L., Scalera, G., Pilone, V., De Sena, G., Quagliariello, V., Iannelli, A., Barbarisi, A.: A comparative study examining the impact of a protein-enriched vs normal protein postoperative diet on body composition and resting metabolic rate in obese patients after sleeve gastrectomy. Obes. Surg. **27**(4), 881–888 (2017)
15. https://medshop.pl/ioi-353-analizator-skladu-ciala-z-oprogramowaniem-jawon-medical.html. Accessed 2 Oct 2019
16. Soares de Araujo, C.G.: Flexitest: an innovative flexibility assessment method. J. Hum. Kinet 72 (2004). https://doi.org/10.2310/6640.2005.00036
17. IPAQ: Guidelines for the data processing and analysis of the International Physical. Activity Questionnaire. https://docs.google.com/viewer?a=v&pid=sites&srcid=ZGVmYXVsdGRvbWFpbnx0aGVpcGFxfGd4OjE0NDgxMDk3NDU1YWRlZTM. Accesed 2 October 2019
18. Zuzda, J.G., Latosiewicz, R., Augustyńska, B.: Risk assessment and level of physical activity of students in Poland. Phys. Educ. Stud. **21**(4), 193–199 (2017). https://doi.org/10.15561/20755279.2017.0408
19. Charkoudian, N.: Skin blood flow in adult human thermoregulation: how it works, when it does not, and why. Mayo Clin. Proc. **7**(603–612), 53–62 (2003)
20. Tanda, G.: The use of infrared thermography to detect the skin temperature response to physical activity. In: 33rd UIT (Italian Union of Thermo-Fluid-Dynamics) Heat Transfer Conference IOP Publishing Journal of Physics: Conference Series, vol. 655 (2015)
21. Chudecka, M., Lubkowska, A.: The use of thermal imaging to evaluate body temperature changes of athletes during training and a study on the impact of physiological and morphological factors on skin temperature. Hum. Mov. Sci. **13**, 33–39 (2012)
22. Zontak, A.: Dynamic thermography: analysis of hand temperature during exercise. Ann. Biomed. Eng. **26**, 988–993 (1998)
23. Al-Nakhli, H.H., Petrofsky, J.S., Laymon, M.S., Berk, L.S.: The use of thermal infra-red imaging to detect delayed onset muscle soreness. J. Vis. Exp. **22**, 59–62 (2012)
24. Clark, R.P., Mullan, B.J., Pugh, L.G.: Skin temperature during running–a study using infrared color thermograph. J. Physiol. **267**, 53–62 (1997)

A Framework for Modeling and Efficacy Evaluation of Treatment of Cancer with Metastasis

Jaroslaw Smieja(✉), Krzysztof Psiuk-Maksymowicz, and Andrzej Swierniak

Silesian University of Technology, Akademicka 16, 44-100 Gliwice, Poland
Jaroslaw.Smieja@polsl.pl

Abstract. The paper is concerned with modeling of tumor growth, taking into account chemo- and radiotherapy and metastasis. The model consists of two compartments, one representing the primary tumor, sensitive to chemo- and radiotherapy and another, representing metastatic tumor. A population of virtual patients is created by sampling model parameters. Kaplan-Meier survival curves are used to analyze the influence of model parameters on treatment efficacy, with one standard treatment protocol.

Keywords: Chemotherapy · Radiotherapy · Metastasis · Modeling

1 Introduction

Lung cancer is one of the most commonly diagnosed cancer and is the leading cause of cancer-related deaths [1] The most common histological subtype is non-small-cell lung carcinoma (NSCLC), accounting for 85% of all lung cancer cases [2]. Advanced NSCLC is more likely to metastasize, leading to severe symptoms and a decrease in overall survival. The presence of distant metastases is one of the most predictive factors of poor prognosis [3].

On the other hand, while mathematical models have been used to analyze cancer growth and treatment outcome for quite a long time (e.g., [4,5] and references therein) to provide insights into treatment efficacy, relatively few of them take metastasis into account. In this work we propose a framework for analysis of primary and secondary tumors, taking into account heterogeneity and various levels of therapy resistance of metastatic tumors. Information gained from massive computational analysis of a large pool of virtual patients may prove to be valuable for clinicians, allowing them to fine-tune therapy protocols.

The authors were supported by NCN grant (National Science Centre, Poland) DEC–2020/37/B/ST6/01959 and SUT statutory research funds. Calculations were partially performed on the Ziemowit computational cluster (http://www.ziemowit.hpc.polsl.pl) created in the POIG.02.01.00-00-166/08 project (BIO-FARMA) and expanded in the POIG.02.03.01-00-040/13 project (Syscancer) and PBS3/B3/32/2015 project (Biotest).

D. G. Pijanowska et al. (eds.), *Biocybernetics and Biomedical Engineering – Current Trends and Challenges*, Lecture Notes in Networks and Systems 293,
https://doi.org/10.1007/978-3-030-83704-4_9

2 Modeling of Tumor Growth and Metastasis

The general idea underlying the model is presented in Fig. 1. Initially, the tumor grows in a primary site and its cells do not have the ability to metastasize. They are assumed to be sensitive to chemo- and radiotherapy. After the primary tumor reaches certain size, the cells can spread to other organs, leading to appearance of a new lung cancer cell in one of metastasis sites. Gaining metastatic potential requires a series of mutations, which may also increase cells growth potential or their resistance to chemotherapeutic agents. Though these mutation events are not included in the proposed model explicitly, they are indirectly represented by both different parameters corresponding to the tumor response to chemotherapy for the primary and secondary tumor, and a random time instant, at which the metastasis is initiated. As a result, the model introduced in this paper considers the metastasis as a single step process, after which both tumors—primary and secondary—grow independently (with a possible flow from primary to metastatic compartment). However, it can be easily changed to take into account metastasis of cells of various mutation levels, colonizing brain, liver or bones, characteristic for NSCLC metastasis—it would require increasing the number of compartments. Such branching processes-based approach, is one of the most common ones, useful also for mathematical modeling of local metastases, as shown in [6] by modeling metastases in pancreatic cancer.

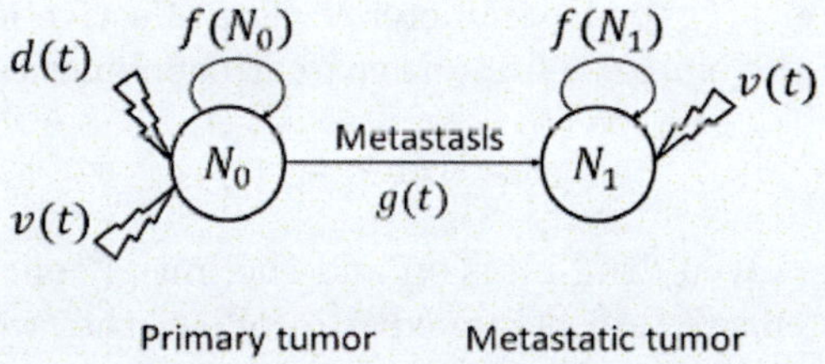

Fig. 1 A schematic illustration of the model.

The minimal model consists of two ordinary differential equations, describing growth of population of cells constituting a primary and secondary tumor, whose size is given by the number of cells, denoted by N_0, and N_1, respectively. Though metastatic cancer may grow in a number of sites, at the same time, we assume the simplest case, in which only one organ has been invaded. Moreover, we do not consider a spatial model and, therefore, even if there are more distant loci, the metastatic tumor is represented by a single compartment. Then, the model is described by the following equations:

$$\begin{aligned} \dot{N}_0 &= f(N_0) - \gamma v(t) N_0 - (\alpha d(t) + \beta d^2(t)) N_0 - g(t) \\ \dot{N}_1 &= g(t) + \epsilon f(N_1) - \gamma_1 v(t) N_1 \end{aligned} \tag{1}$$

with initial conditions

$$\begin{aligned} N_0(0) &= N_{in} \\ N_1(0) &= 0 \end{aligned} \tag{2}$$

The function $g(t)$ represents the migration of cells from the primary to secondary tumor. If the metastasis is considered to be a singular event, taking place at time t_m, then $g(t) = N_s\delta(t-t_m)$, where N_s represent the number of cells that migrated from the primary site and initiated tumor growth at the secondary site and $\delta(.)$ is the Kronecker delta. Alternatively, if a constant flow of migrating cells is considered, $g(t) = N_s\mathcal{H}(t-t_m)$, where $\mathcal{H}(.)$ is the Heaviside function. Simulation results, shown in this paper are based on the former form of the function.

The function $f(N_i)$ represents the tumor growth term. It may assume different forms, thus affecting both population dynamics and the form of optimal treatment schedules (see, e.g. [10]). In this work, the Gompertz form has been chosen:

$$f(N_i) = -\rho_g N_i \ln \frac{N_i}{K_i} \tag{3}$$

where K_0 and K_1 represent so-called carrying capacities for the primary and secondary tumors, respectively.

The model assumes that chemotherapy kills the tumor cells according to the log-kill hypothesis [11], hence the kill rate is proportional to chemotherapeutic agent concentration $v(t)$ with parameter γ. The effective drug concentration is assumed to follow the simplest first-order pharmacokinetics:

$$\dot{v} = -\lambda v + k_v c(t) \tag{4}$$

with $c(t)$ being the given drug infusion and the initial concentration $v(0) = 0$.

Radiotherapy effects are assumed to follow the well-established linerquadratic (LQ) model [12], with $d(t)$ denoting irradiation dose at time t and α, β being LQ model parameters. The model assumes that the metastatic tumor can be affected by chemotherapy only. A separate parameter γ_1 represents effectiveness of the drug on the metastatic cancer cells (if metastatic cancer is at least partially drug-resistant, then $\gamma_1 < \gamma$).

In clinical practice, radiotherapy is administered as a series of impulses. However, these may be replaced by a biologically equivalent dose, constant during the radiotherapy period, yielding the same simulation results. Therefore, in this work $d(t)$ is a constant over the period of non-zero radiation therapy.

The parameter t_m represents time needed for the primary tumor to gain metastatic potential with N_s cells initiating metastatic tumor growth. To take into account potentially more aggressive metastatic tumor growth, as compared to the primary tumor, the parameter ϵ has been introduced.

A sample growth dynamics without therapy actions is shown in Fig. 2.

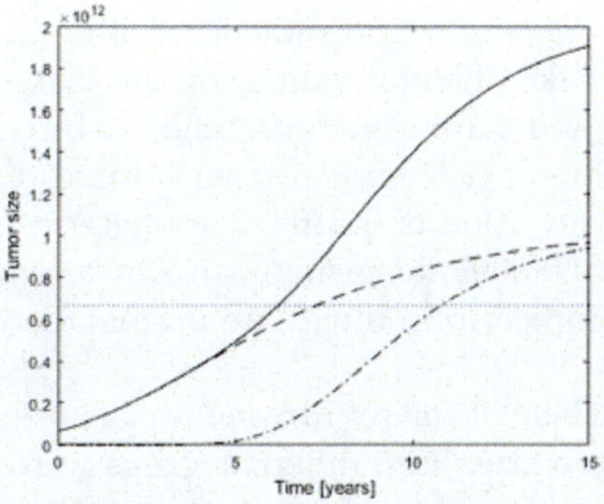

Fig. 2 Growth of primary (dashed line) and secondary (dotted line) tumor, together with the total tumor size (solid line)

3 The Framework for Analysis

Kaplan-Meier survival curves [9] have been chosen as the main tool facilitating comparison of numerical results. A virtual pool of patients have been created, whose tumor growth and response to treatment is modeled according to (1). To take into account inter-patient tumor heterogeneity, model parameters were sampled from either normal, uniform or exponential distributions, as indicated in the Table 1.

Initial size of tumor is sampled as the diameter d_{N^0} of the primary tumor at the moment of detection. For the sake of this work, it was assumed following [10], that its diameter at the beginning of treatment is equal to 6 cm. Death of the patient is assumed to take place after the diameter of the tumor d_{N^d} reaches 13 cm. The "carrying capacity" parameter d_K, describing possible tumor diameter in the given amount of nutrition and oxygen, is assumed to be equal to 30 cm [10]. It has to be noted that although the K value is described as the maximal size of the tumor in given environmental conditions, it does not directly translate into biological size. The value is rather serving the purpose of throttling the growth of the tumor.

The tumor is assumed to be of spherical shape, thus the volumes N^0, K, N^d are calculated as ones of spheres with diameters as described previously. The density of cancer cells per cm^2 is assumed to be equal to $5.8 \cdot 10^8$ per cm^2 [7]. The density should not, however, have any impact on the overall results, as it simply scales volume into the number of cells equally. The death condition is recalculated into the sum of size of both populations.

The time of metastasis appearance t_m, at which the first group of cells settles in the new location, is given by $t_1 + t_2$, where t_1 is the moment the tumor reaches the size after which metastasis can be initiated and t_2 is sampled from an exponential distribution, thus is characterized by arbitrarily chosen mean value μ.

The parameter ρ, determining the tumor growth rate is sampled from a normal distribution with a mean of $7.00 \cdot 10^{-5}$ and a standard deviation of

$7.23 \cdot 10^{-3}$. Parameter *epsilon* responsible for scaling the metastasis growth rate has been assumed to take different values for the sake of further analysis.

The parameters α and β are usually assumed to be connected by the following relation: $\beta = \frac{\alpha}{10}$. Value of α is chosen from normal distribution with mean of 0.0398 and standard deviation of 0.0168. Parameter γ is a constant. Its value is assumed to be equal to 0.0093, following [14]. Due to correlation of parameters ρ and α [7], they are sampled from biviariate normal distribution with correlation coefficient 0.87.

The drug degradation parameter *lambda* is calculated using the drug half-life and the simulation step size. Two different drugs were assumed to be administered, one being cisplatin in dose of 5 mg/m^2 and the other vinblastine in dose of 100 mg/m^2, according to the standard treatment protocol [8] (Fig. 3).

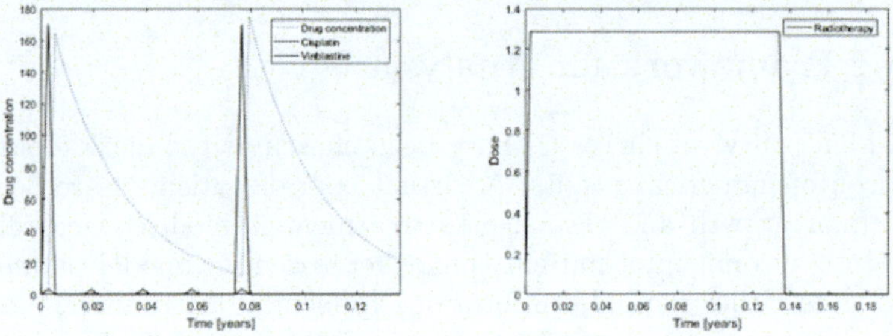

(a) (b)

Fig. 3 a Chemotherapy and **b** Radiotherapy protocols

Table 1 Parameters, distributions and values

Parameter	Distribution	Value
ρ_g	Normal	$7.00 \cdot 10^{-5}$ (μ), $7.23 \cdot 10^{-3}$ (σ)
ρ_e	Normal	$\zeta \cdot \rho_g$
α	Normal	0.0398 (μ), 0.0168 (σ)
β	Normal	$\frac{\alpha}{10}$
γ	Uniform	0.0093
ϵ	Uniform	$\langle 1; 2 \rangle$
η	Uniform	$\langle 1; 1.3 \rangle$
d_{N^0}	Uniform	6 [cm]
d_K	Uniform	30 [cm]
d_{N^d}	Uniform	13 [cm]
t_m	Exponential	$\langle 20; 100 \rangle$
λ_c	Constant	0.1

4 Results and Discussion

4.1 Survival Curves for Model Without Treatment

First, the difference between survival function for patients with and without metastasis is compared (Fig. 4), assuming that the primary and metastatic tumors exhibit exactly the same dynamics (with the same parameters).

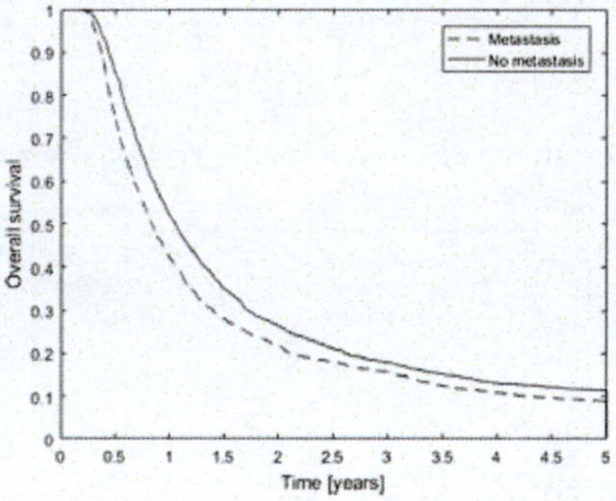

Fig. 4 Kaplan-Meier survival curves for non-metastatic and metastatic cancer without treatment

The survival of patients with secondary tumor is smaller than of those with only one cancer site. The difference is not large, though (most visible 1 to 1.5 years after the detection of the primary tumor). Such small difference is due to the fact that the metastasis was modeled as a singular event only. taking into account that there is no influx of new cells from the primary tumor afterwards and only in the metastatic site the (partial) drug resistance is observed. Moreover, it was assumed that the same parameters define growth of both tumors. Therefore, to make computational analysis more relevant, either constant flow from primary to secondary compartment or multiple discrete-time metastatic events should be included in the model.

Subsequently, the influence of aggressiveness of the metastatic tumor, defined by the ϵ parameter was checked (Fig. 5). The respective survival percentages at the end of the observation period were 10.4, 8.5 and 7.2%. The difference is greater between factor 1 and 1.5 than between 1.5 and 2. The change in survival rates at the final time is not linearly dependent on this parameter, due to nonlinear form of Gompertz growth function.

4.2 Survival Curves for the Model with Treatment

Since the metastatic tumor is not affected by the radiotherapy at all, and may exhibit partial resistance to chemotherapy, significant differences between survival curves obtained for models with and without metastasis should be expected. Indeed, the difference proved to be significant (Fig. 6).

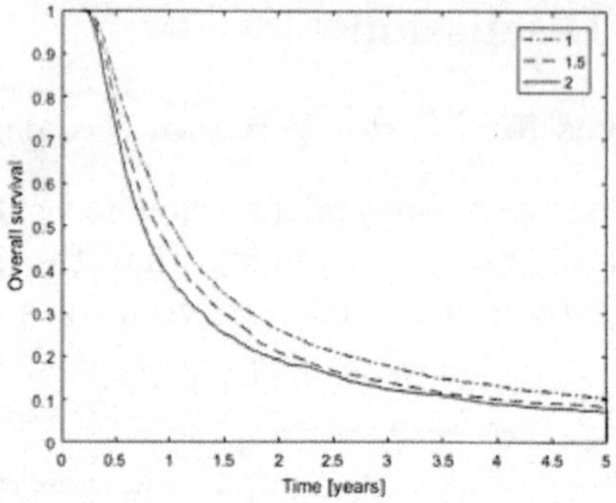

Fig. 5 Kaplan-Meier survival curves for metastasis with variable parameter ϵ

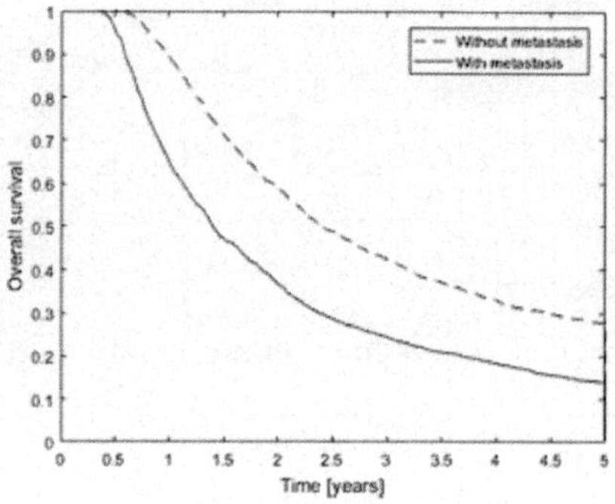

Fig. 6 Growth of tumor with and without metastasis, with treatment

The survival curves for both models branch out from the moment of tumor detection. Only during the first six months they are similar, which can be explained by the fact that initially it is the primary tumor that affects the patient state.

4.3 Impact of Time of Metastasis Occurrence on the Survival Curves

The time at which metastasis occurs is potentially a significant factor affecting survival curves. Therefore, for the next numerical experiment different mean values of the distributions, used to sample that moment, were assumed: 20, 50 and 100 (Fig. 7). Though the two latter are placed further apart, the difference in the shape of survival curves is more prominent between the two former. The $\mu = 20$ results in survival of 10.5%, $\mu = 50$ of 9%, $\mu = 100$ of 7.5% at the final time.

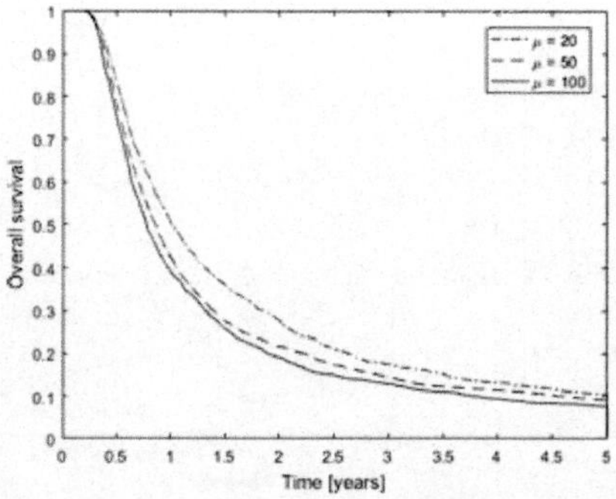

Fig. 7 Survival curves for metastasis with different seeding time distributions ϵ without treatment applied

4.4 Impact of Environmental Limitations on the Survival Curves

One of the important factors affecting tumor growth is availability of nutrients needed by the expanding population of cells. After the primary tumor reaches a certain size, the vascular network might not be capable of supporting its further growth. This may cause the carrying capacity K to be smaller than the actual death condition value. However, if metastatic tumor is taken into account, the question of how to define the death condition properly arises.

For sake of this analysis, the carrying capacity was assumed to be equal to 12 cm, both in the primary and secondary site. In the model compared $K =$ 15 cm, so it is larger than the death condition.

The size of 12 cm was chosen arbitrarily, as it is smaller than the death condition, but is sufficiently large that two tumors of that diameter would fulfill the death condition. This would not be the case with the diameter of i.e. 10 cm, as a total size of both the primary and the metastatic tumors would still be smaller than the death condition, thus no deaths would be registered at all.

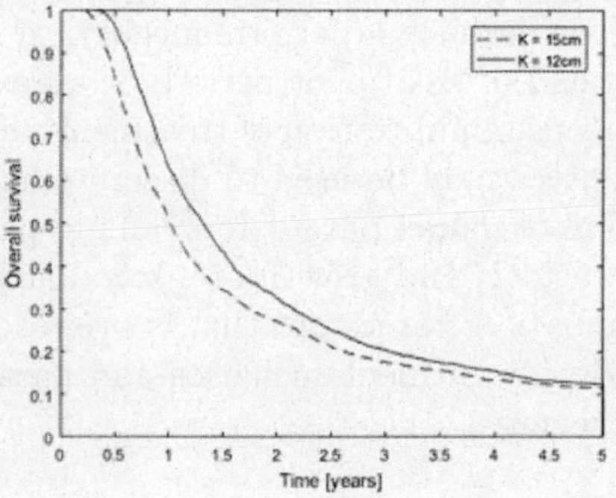

Fig. 8 Survival curves for model with metastasis, without treatment

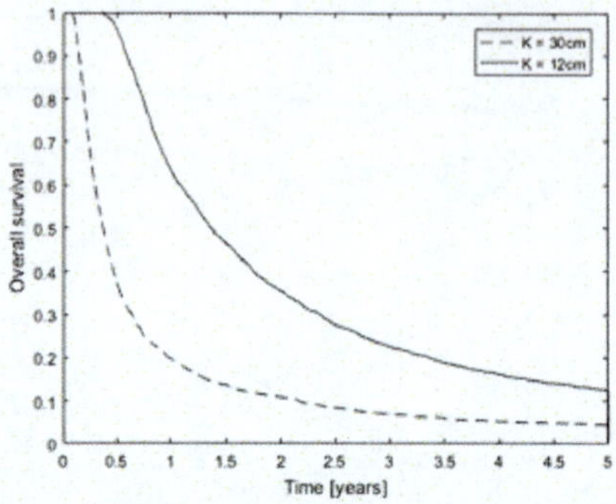

Fig. 9 Survival curves for model with metastasis, without treatment

The result of the simulations is presented in Fig. 8. The result is not quite as one could expect. There is almost no difference between survival at the end of the observation period. For $K = 12$ cm the survival is equal to 12.3% and for $K = 15$ cm it is 11.4%. The biggest difference is visible in the middle of the curve, reaching close to 13% after approximately 1 year. That indicates that the carrying capacity and the threshold determining the death condition and should be set at a lower level, if metastasis is taken into account. The specific value of the threshold should be estimated with patients' survival data, as it is affected by e.g. the number of metastatic sites, which in the model are combined into one. The sample result, comparing the impact of reducing the carrying capacity to 12 cm is shown in Fig. 9. Here the survival at the end of the observation period is much lower.

5 Conclusions

Mathematical modeling of tumor growth and its response to anticancer treatment facilitates prediction of treatment outcome both for an individual patient, and for a population of patients. In the former case, model simulation for various treatment protocols makes it possible to choose the one that is the best for that patient, taking into account a performance index, defined usually as either the tumor size at the end of treatment period, or an average tumor size in this period, with simultaneous minimization of treatment negative side effects. In the latter case, survival curves may be used to discern efficacy of various protocols, assuming a distribution of model parameters among patient population.

In both cases, metastasis and growth of a secondary tumor should be taken into account. The simulation framework that is proposed in this work, allows to take into account various treatment scenarios and parameters, as well as inter-patient tumor heterogeneity.

Acknowledgements. The Authors would like to thank prof. Suwinski from the Maria Sklodowska-Curie Institute - Oncology Centre (MSCI), branch in Gliwice, for valuable comments on clinical applicability of the results obtained from modeling, and Ms. Anna Tasarz, who provided simulation results.

References

1. Inamura, K.: Lung cancer: understanding its molecular pathology and the 2015 WHO classification. Front. Oncol. **7**, 193 (2017). https://doi.org/10.3389/fonc.2017.00193
2. Dela Cruz, C.S., Tanoue, L.T., Matthay, R.A.: Lung cancer: epidemiology, etiology, and prevention. Clin, Chest Med. **32**, 605–644 (2011) https://doi.org/10.1016/j.ccm.2011.09.001
3. Popper, H.H.: Progression and metastasis of lung cancer. Cancer Metast. Rev. **35**(1), 75–91 (2016). https://doi.org/10.1007/s10555-016-9618-0
4. Swierniak, A., Kimmel, M., Smieja, J., Puszynski, K., Psiuk−Maksymowicz, K.: System Engineering Approach to Planning Anticancer Therapies, Springer International Publishing Switzerland, Cham (2016). https://doi.org/10.1007/978-3-319-28095-0
5. Schaettler, H., Ledzewicz, U.: Optimal Control for Mathematical Models of Cancer Therapies. An Application of Geometric Methods, Springer Science + Busuness Media, New York (2015). https://doi.org/10.1007/978-1-4939-2972-6
6. Haeno, H., Gonen, M., Davis, M.B., Herman, J.M., Iacobuzio-Donahue, C.A., Michor, F.: Computational modeling of pancreatic cancer reveals kinetics of metastasis suggesting optimum treatment strategies. Cell **148**, 362–375 (2012). https://doi.org/10.1016/j.cell.2011.11.060
7. Geng, C., Paganetti, H., Grassberger, C.: Prediction of treatment response for combined chemo- and radiation therapy for non-small cell lung cancer patients using a bio-mathematical model. Sci. Rep. **7** (2017)
8. Curran, W.J., Jr., Paulus, R., Langer, C.J., Komaki, R., Lee, J.S., Hauser, S., Movsas, B., Wasserman, T., Rosenthal, S.A., Gore, E., Machtay, M., Sause, W., Cox, J.D.: Sequential vs concurrent chemoradiation for stage III nonsmall cell lung cancer: randomized phase III trial RTOG 9410. JNCI J. Natl. Cancer Inst. **103**(19), 1452–1460 (2011)
9. Dudley, W.N., Wickham, R., Coombs, N.: An Introduction to Survival Statistics: Kaplan-Meier Analysis. J Adv Pract Oncol **7**(1), 91–100 (2016)
10. Dolbniak M., Smieja, J., Swierniak, A.: Structural sensitivity of control models arising in combined chemo-radiotherapy. In: Procceedings of the 23rd International Conference on Methods and Models in Automation and Robotics (MMAR), pp. 339–344 (2018)
11. Skipper, H.E., Schabel, F., Wilcox, W.: Experimental evaluation of potential anticancer agents. xiii. on the criteria and kinetics associated with curability of experimental leukemia. Cancer Chemother. Rep. **35**, 1–111 (1964)
12. Fowler, J.F.: The linear-quadratic formula and progress in fractionated radiotherapy. Br. J. Radiol. **62**, 679–94 (1989)
13. Lee, J.Y., Kim, M.-S., Kim, E.H., Chung, N., Jeong, Y.K.: Retrospective growth kinetics and radiosensitivity analysis of various human xenograft models. Lab. Anim. Res. v. **32**(4), 187–193 (2016). https://doi.org/10.5625/lar.2016.32.4.187
14. Swierniak, A., Smieja, J., Mura, M., Bajger, P.: Modeling and optimization of radio-chemotherapy. In: Current Trends in Biomedical Engineering and Bioimages Analysis, pp. 223–233 (2020). https://doi.org/10.1007/978-3-030-29885-2_20

Hybrid Cardiovascular Simulator - An Application for the Mechanical Assistance by an Intra-aortic Balloon Pump

Raman Pasledni(✉) and Krzysztof Zieliński

Nalecz Institute of Biocybernetics and Biomedical Engineering, Polish Academy of Sciences, Ks. Trojdena 4, 02109 Warsaw, Poland
rpasledni@ibib.waw.pl

Abstract. The United States Food and Drug Administration agency is encouraging to use modeling and simulation in medical device evaluation, in order to reduce reliance on animal models and human data, and reduce costs and speed innovation. To address this challenge various models and simulators have been developed that play the role of artificial/virtual patients. A previously elaborated in IBBE PAS a hybrid (hydro-computer) simulator was adapted to perform simulations of the left ventricular mechanical assistance by the intra-aortic balloon pump (IABP). To implement this, some hardware and software changes were made to the simulator, in particular the following functionalities were added: connecting a physical aorta model; variable aortic compliance by introducing a numerical capacitor into the hybrid circulatory model; a rigid tube model of the aorta with an orifice system to simulate IABP support. Static characteristics were measured for both physical models of the aorta (silicone tubes) and the model with a numerical capacitor. IABP assistance was simulated using the developed aortic model. The preliminary simulation results show that the system is behaving correctly. The effects of IABP mechanical assistance of the left ventricle observed in the simulator are consistent with clinical practice. The application for simulations of IABP mechanical assistance extends the whole IBBE PAS hybrid simulator functionalities in terms of the mechanical heart assistance simulations.

Keywords: Cardiovascular simulator · IABP assistance · Aortic compliance

1 Introduction

The United States Food and Drug Administration (FDA) agency is encouraging to use modeling and simulation in medical device evaluation, in order to reduce reliance on animal models and human data, and reduce costs and speed innovation [1]. To address this challenge many different models and simulators have been developed that play the role of artificial/virtual patients [2, 3]. In the case of cardiovascular system, typical approach is to develop and utilize hydraulic mock-up simulators, allowing the interaction with physical cardiovascular devices (e.g. ventricular assist devices (VADs), heart valves prosthesis

D. G. Pijanowska et al. (eds.), *Biocybernetics and Biomedical Engineering – Current Trends and Challenges*, Lecture Notes in Networks and Systems 293,
https://doi.org/10.1007/978-3-030-83704-4_10

etc.) [3]. An alternative solution for mock-up hydraulic systems are hybrid simulators [4] of the human cardiovascular system, such as developed in the Nalecz Institute of Biocybernetics and Biomedical Engineering Polish Academy of Sciences (IBBE PAS) [5]. This IBBE PAS simulator consists of a numerical model of the cardiovascular system implemented and executed in the real-time system and special physical part that allows connecting and testing cardiovascular assist devices [3, 4].

Intra-aortic balloon pump (IABP) therapy is a widely used intervention for temporary assistance in a variety of clinical situations such as the left ventricular failure or acute myocardial infarction with cardiogenic shock [6]. The effects of IABP are based on the principle of counterpulsation, whereby blood is pumped or displaced out of phase with the normal cardiac cycle. Balloon inflation causes an increment of mean arterial blood pressure and myocardial perfusion. Balloon deflation reduces afterload of the following beat, which causes a reduction of left ventricular wall tension. The overall benefit is an increment of the ratio of myocardial oxygen supply to oxygen demand [7].

There are several studies dedicated to simulations of IABP assistance. Some of them were performed by a mock-up cardiovascular model [8], other on a hybrid (hydro-numerical) model of the cardiovascular system [9, 10]. In these cases, an elastic tube was used as a descending aorta model. However this approach does not allow changes in the compliance of the aorta during simulation. However, there are a number of studies [11, 12] which prove that aortic compliance varies greatly in sick people (it progressively lower in patients with hypertension or subclinical aortic atherosclerosis). Therefore, it is very important to be able to change compliance during simulations.

Therefore, the aim of this paper was to introduce to a previously elaborated hybrid (hydro-computer) cardiovascular simulator, an application to support the use of IABP assistance. Moreover, the ability to change the descending aorta compliance was performed by introducing an additional numerical compliance in the hybrid cardiovascular model.

2 Materials and Methods

2.1 The Hybrid Cardiovascular Simulator

The hybrid cardiovascular simulator, developed in IBBE PAS was used in this study. It consists of two main functional modules: the computer model of the cardiovascular system and the hydro-numerical interface.

The computer model of the cardiovascular system mechanics is based on the model presented in [5, 13]. The heart ventricles systole is described by the time-varying elastance model, while their ventricular filing phase is simulated by the sum of exponential functions [13, 14]. Atriums are described by passive, linear compliances. The systemic and pulmonary circulations are described by 4 Windkessel models for arterial systemic, venous systemic, arterial pulmonary and venous pulmonary circulation, respectively. Each Windkessel model reproduces resistive, inertial and elastic properties of the simulated vessels.

The hydro-numerical interface consists of 4 sections. Each section, called "impedance transformer" (TR) [3], converts bidirectionally pressure and flow signals from the physical (hydraulic) domain to the numerical one. Thanks to it, the mutual interaction of physical assist devices and the computer circulatory model is possible. 4 sections allow connection of physical medical assist device or prosthesis up to 4 points to the numerical model. The simulator is also equipped with a rectangular chamber with rigid walls, into which you can install various tubes that act as the descending aorta (see Fig. 1).

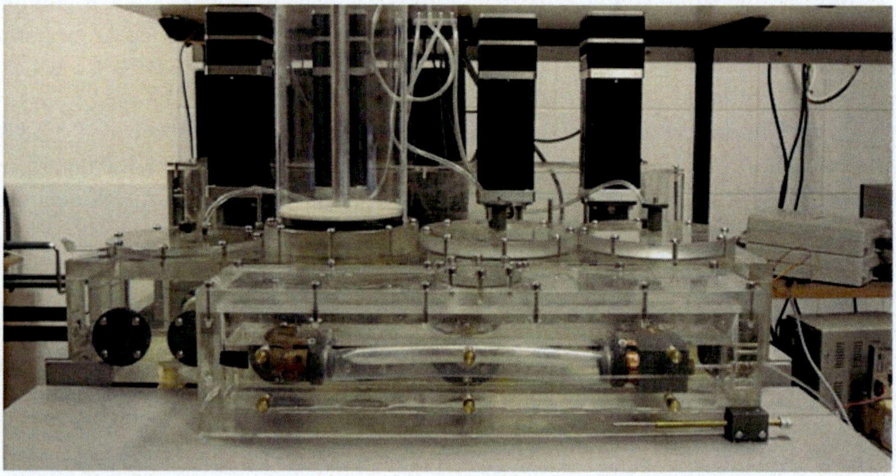

Fig. 1 Image of a hybrid cardiovascular simulator with an additional chamber to accommodate the model of the aorta

The hybrid interface as well as hybrid modeling approach was described in more details in [3–5].

2.2 Static Characteristics of Descending Aorta Models

Two silicone tubes were used as physical models of the descending aorta. They have the same geometric dimensions: 25 cm in length and 2.7 cm diameter but different wall thickness (0.45 and 1 mm).

Experimental setup for measuring static characteristics of physical descending aorta model is shown in Fig. 2a. The fluid was injected into the aorta model in 10 ml increments using a syringe. The pressure inside the aorta model was measured using a PS10063-002 sensor (Schaevitz, UK) with measuring range up to 0.35 bar.

To measure the static characteristic of the descending aorta model with a numerical capacitor the laboratory setup shown in Fig. 2b was used. The hybrid impedance transformer (TR) provides an interface between numerical and physical sections. It is used to generate a controlled fluid flow in accordance with a mathematical (numerical) model, described in the next paragraph.

To create a numerical model of the descending aorta (with numerical capacitor), an electro-hydraulic analogy was used.

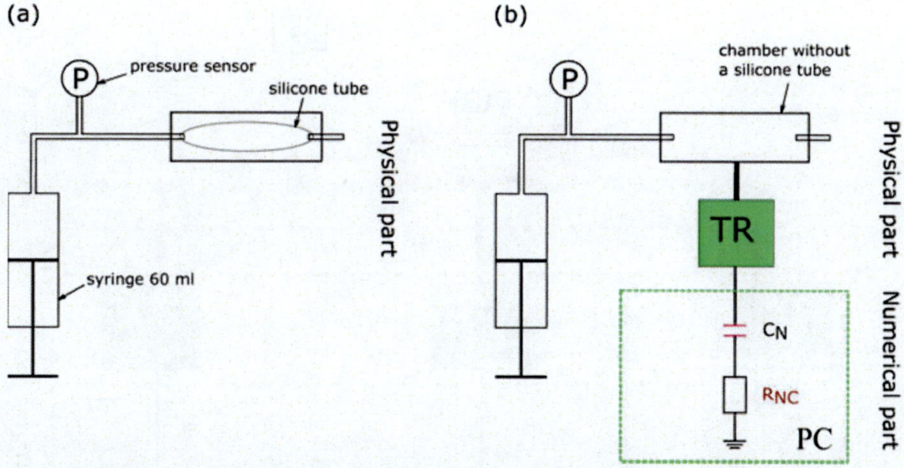

Fig. 2 Experimental setup for measuring static characteristics of the aorta: **a** for physical models; **b** for a model with a numerical capacitor. TR—impedance transformer, C_N—numerical capacitor, R_{NC}—numerical resistor

The mathematical model characterizing aortic viscoelasticity can be represented by the following expressions:

$$q \times R_{NC} + P_{CN} - P_S = 0 \quad (1)$$

$$q = C_N \times \frac{dP_{CN}}{dt} \quad (2)$$

where q is the fluid flow rate; P_{CN} is the pressure in the chamber caused by the introduction of a numerical capacitor into the system; P_S is source pressure; C_N, R_{NC} are the numerical capacitor and the numerical resistor, respectively.

Initial measurements were conducted without a numerical capacitor and a physical aorta to estimate the own compliance of the system (C_{int}). The next step was to measure and obtain static characteristics for different values of numerical capacitor C_N as shown in Fig. 2b.

2.3 The Hybrid Simulator of the Cardiovascular System Development for IABP Assistance

In this study, the descending aorta model was integrated with the rest of the numerical model of the circulatory system. A diagram for studying dynamic characteristics of physical aortic model with a numeric capacitor and for studying the effect of IABP assistance is shown in Fig. 3.

G is an electro-pneumatic controller. TR1, TR2, TR3 are impedance transformers. TR1 is used to connect the physical aortic segment before IABP with the computer model of the left ventricle, aortic arch and coronary circulation, while TR2 is used to connect aortic segment after IABP with a computer model of systemic circulation, and TR3 is used to simulate the viscoelastic properties of the descending aorta model. As the coronary circulation model a simple Windkessel model [15] was used.

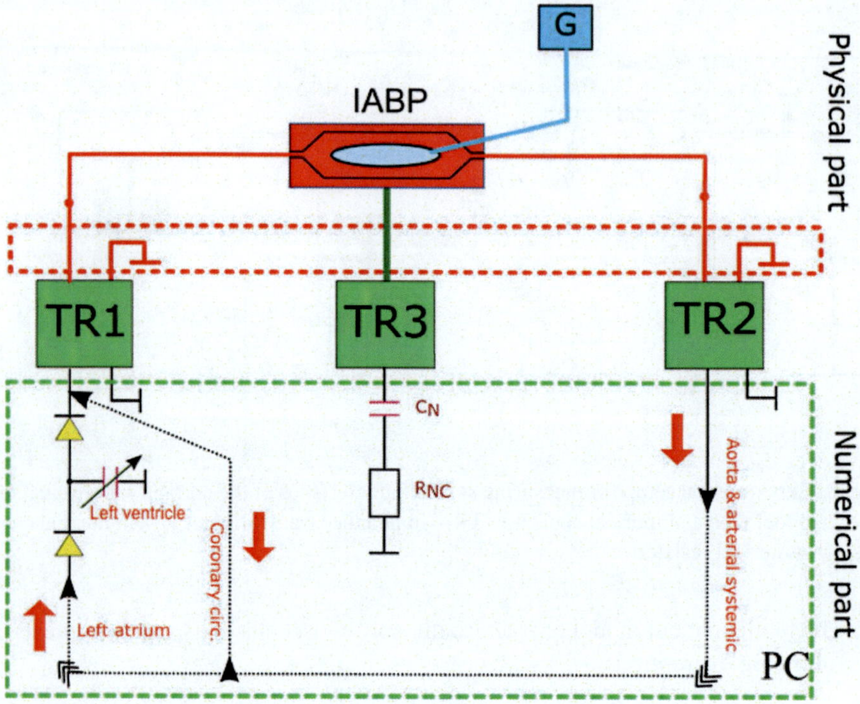

Fig. 3 Diagram for studying dynamic characteristics of physical aortic model with C_N and for studying effects of intra-aortic balloon pump (IABP) assistance

2.4 The Hybrid Cardiovascular Simulator Configuration for IABP Assistance

In this study, two physical models of the descending aorta were used: the first is a rigid tube with holes; the second is a silicon tube with a wall thickness of 0.45 mm. A 50 ml IABP is placed inside the physical descending aorta model.

The rigid tube with holes was designed specially for cooperation with numerical model of the viscoelasticity of the descending aorta.

A dedicated, electro-pneumatic controller (*G*) was used for the balloon inflation and deflation. Its work is synchronized with the left ventricular model (quasi ECG signal indicating new cardiac cycle in the model). After a certain time delay T_{del} from the beginning of left ventricular systole, the controllers software generates a signal, its rising edge triggers the IABP inflation. The balloons inflation continues until the end of diastole, and then it is deflated at the beginning of the next cardiac cycle in the model. The time delay T_{del} is set manually using dialog box control in the user interface.

For the descending aorta model as a rigid tube with holes, the exemplary left ventricular failure was simulated and then next simulations were carried out for four C_N values (0, 0.5, 1.0, 1.5), with the time delay $T_{del} = 0.3$ s.

For the descending aorta model as a silicone tube, T_{del} was varied from 0.2 to 0.5 s with a step of 0.05 s. These simulations were performed with the following values of the parameters characterizing the cardiovascular model (Table 1).

The IABP performance was tested for set inflation pressure $P_{inf} = 150$ mmHg.

Table 1 List of cardiovascular parameters used in the simulations

Cardiovascular parameters	Abbreviations	Units	Values
Heart rate	HR	beats/min	60
Left/right ventricular rest volume	V0l/V0r	cm^3	5/5
Left/right maximum time-varying elastance	Emaxl/Emaxr	$mmHg/cm^3$	1/1.01
Left/right atrial compliance	Cla/Cra	$cm^3/mmHg$	30/30
Left/right ventricular input resistance	Rli/Rri	$mmHg{\cdot}s/cm^3$	30/0.0215
Left/right ventricular output resistance	Rlo/Rro	$mmHg{\cdot}s/cm^3$	10/0.0215
Systemic/pulmonary characteristic Resistance	Rcs/Rcp	$g{\cdot}s/cm^4$	200/0.005
Systemic/pulmonary arterial compliance	Cas/Cap	$cm^3/mmHg$	1.5/1
Systemic/pulmonary arterial resistance	Ras/Rap	$mmHg{\cdot}s/cm^3$	1/0.1
Systemic/pulmonary venous compliance	Cvs/Cvp	$cm^3/mmHg$	83/6
Systemic/pulmonary venous resistance	Rvs/Rvp	$mmHg{\cdot}s/cm^3$	0.007/0.005
Coronary resistance	Rcor	$mmHg{\cdot}s/cm^3$	500

In order to evaluate the behavior of the descending aorta model with a numerical capacitor C_N and to assess the effectiveness of IABP support, simulations were performed with and without IABP assistance. For simulations performed with a silicone model of the aorta, the following parameters were calculated and analyzed: cardiac output (*CO*), left ventricular end-systolic pressure (*LVESP*), left ventricular end-diastolic pressure (*LVEDP*), mean arterial systemic pressure (*MASP*), arterial end-systolic pressure (*AESP*), arterial end-diastolic pressure (*AEDP*) and mean coronary flow (MCF).

3 Results and Discussion

3.1 Static Characteristics of Physical and Numerical Aortic Models

The experimental data were used to obtain static characteristics of the model of descending aorta (silicone tubes with a wall thickness of 0.45 and 1 mm) and models with a numeric capacitor C_N (see Fig. 4).

The characteristics for the numerical compliances are linear (the correlation coefficients greater than 0.999) which shows that the transformation of the numerical capacitor C_N is working correctly. However, the actual compliance values resulting from the slope of these curves are greater than the set values. This is because the internal compliance C_{int} is sequentially added to the simulated compliance C_N. The resulting compliance is also affected by the pressure regulator error in the pressure source used in TR3.

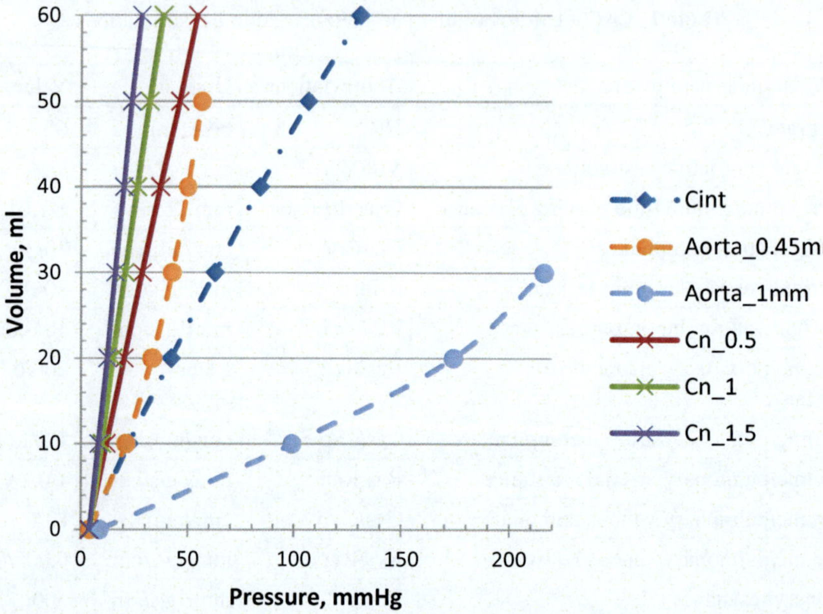

Fig. 4 Static characteristics (SC) of descending aorta models. Where C_{int} is the SC of the system with its own compliance (without C_N and physical aortic model); Aorta_0.45 mm, Aorta_1 mm are SC of physical aortic models; Cn_0.5…Cn_1.5 are SC of aortic models with numerical capacitors

3.2 The Results of Using Rigid Tube as the Descending Aorta Model with a Numerical Capacitor during IABP Assistance

The figures below (Figs. 5, 6 and 7) show the simulation results for different values of the numerical capacitor C_N during IABP assistance. Changes in left ventricular pressure (*Plv*) and aortic pressure (*PAS*) were taken to assess the influence of the numerical C_N on the total elasticity of the descending aorta model in the hybrid simulator. Moreover, each figure shows the pressure changes in the case of IABP assistance (*Plv_with IABP*, *PAS_with IABP*) and without it (*Plv_without IABP*, *PAS_without IABP*).

The aortic pressure shows the influence of the numerical capacitor on the elastic properties of the hybrid descending aorta model in which the IABP is placed. The aortic compliance is the most visible by the difference between maximal and minimal PAS pressure. Higher compliance then lower values of this difference what is observed in the simulation results.

3.3 The Results of Using Silicon Tube as the Descending Aorta Model During IABP Assistance

Table 2 shows the simulation results for different support times T_{del} during IABP assistance.

Based on the collected data, it can be said that the system was behaving in accordance what is observed in clinical practice. The effects of the left ventricular assistance by IABP is observed. The best results are obtained for T_{del} times of 0.35 and 0.4 s, since

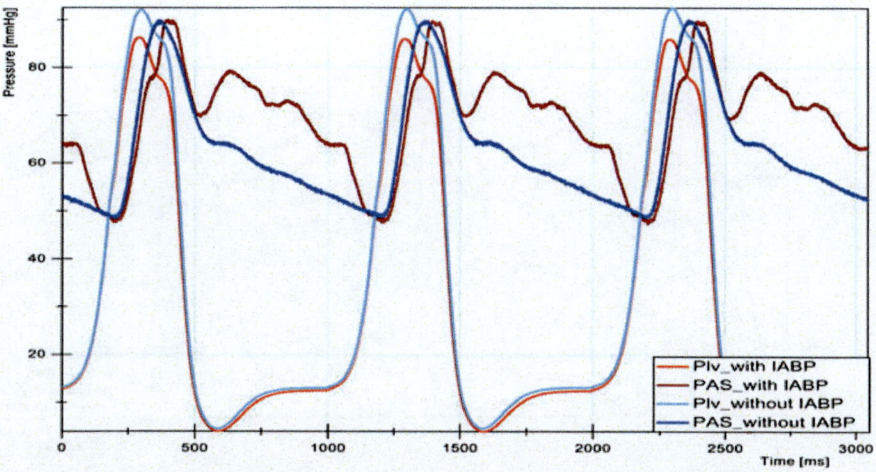

Fig. 5 Left ventricular pressure and aortic pressure with and without IAPB assistance, for $C_N =$ 0 mL/mmHg

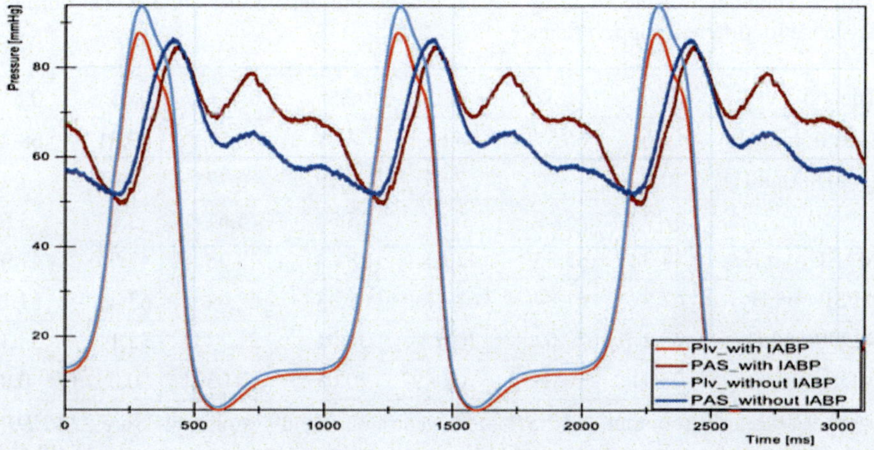

Fig. 6 Left ventricular pressure and aortic pressure with and without IAPB assistance, for $C_N =$ 0.5 mL/mmHg

the parameters MCF and MASP have the highest values and the LVESP values are also minimal (afterload reducing is a direct goal of IABP assistance). For $T_{del} = 0.4$ s we also have the lowest LVEDP which provides the best preload reduction of the left ventricle. The IABP inflates and deflates by means of a counter pulsation. It deflates during heart systole and inflates during its diastole. The simulations confirmed this. Too early inflation (low T_{del} values) causes that the IABP inflation is started during the systole, what worsening the IABP assistance effectiveness. Too high T_{del} means that the inflation time is too short. The perfect deflation time is at the beginning of the diastole what was shown in simulations.

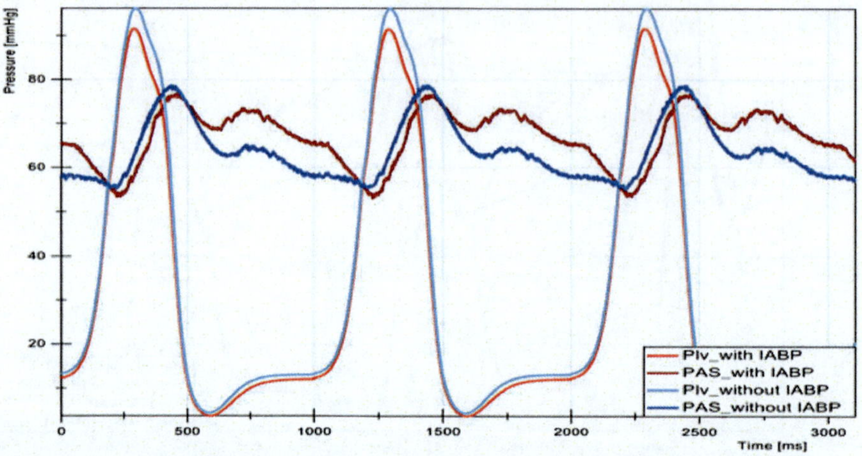

Fig. 7 Left ventricular pressure and aortic pressure with and without IAPB assistance, for C_N = 1.5 mL/mmHg

Table 2 Simulation results of using silicon tube as the descending aorta model during IABP assistance for different support times

Tdel (s)	0.20	0.25	0.3	0.35	0.4	0.45	0.5
LVESP (mmHg)	69.6	66.57	66.65	66.61	66.57	67.51	68.26
LVEDP (mmHg)	10	9.71	7.93	7.19	6.77	7.06	7.22
CO (l/min)	5.17	5.19	5.48	5.64	5.66	5.67	5.63
MASP (mmHg)	44.38	45.35	47.63	48.64	48.08	47.9	47.91
AESP (mmHg)	67.3	61.89	62.17	62.33	62.49	63.45	64.13
AEDP (mmHg)	28.2	29.02	30.5	31.04	31.37	32.11	33.34
MCF (cm^3/s)	0.0411	0.0420	0.0497	0.0525	0.0521	0.0515	0.0509

where ***CO***—cardiac output, ***LVESP***—left ventricular end-systolic pressure, ***LVEDP***—left ventricular end-diastolic pressure, ***MASP***—mean arterial systemic pressure, ***AESP***—arterial end-systolic pressure, ***AEDP***—arterial end-diastolic pressure, ***MCF***—mean coronary flow

4 Conclusions

According to the preliminary tests, the presented idea of the numerical model of the viscoelastic properties of the descending aorta, transformed into the hydraulic form by the impedance transformer and the rigid tube with holes, works correctly. However further studies are needed to improve its performance. The preliminary simulation results of the IABP mechanical assistance by means of physical balloon proved that the whole system also works correctly.

The application for simulations of IABP mechanical assistance extends the whole IBBE PAS hybrid simulator functionalities in terms of the mechanical heart assistance simulations. This application can also be used to simulate the hypertension or aortic atherosclerosis, as the values of aortic compliance in the physical part of the hybrid simulator can be modified during the simulations.

Acknowledgements. We would like to express our special thanks to PROMED S.A. company that gave us a great opportunity to test the IABP (ARROW ® Product)

Funding. The study received financial support from the Nalecz Institute of Biocybernetics and Biomedical Engineering, Polish Academy of Sciences.

Ethical Approval. This article does not contain any studies with human participants or animals performed by any of the authors.

References

1. CDRH Regulatory Science Priorities. https://www.fda.gov/medical-devices/science-and-research-medical-devices/cdrh-regulatory-science-priorities. Accessed 20 Mar 2021
2. Ninomiya, S., Tokaji, M., Tokumine, A., Kurosaki, T.: Virtual patient simulator for the perfusion resource management drill. J. Extra Corpor. Technol. **41**(4), 206–212 (2009)
3. Fresiello, L., Zieliński, K.: Hemodynamic modelling and simulations for mechanical circulatory support. In: Karimov J., Fukamachi K., Starling R. (eds.) Mechanical Support for Heart Failure. Springer, Cham (2020). https://doi.org/10.1007/978-3-030-47809-4_26
4. Zieliński, K., Darowski, M., Kozarski, M., et al.: The need for hybrid modeling in analysis of cardiovascular and respiratory support. Int. J. Artif. Organs. **39**(6), 265–271 (2016)
5. Darowski, M., Kozarski, M., Ferrari, G., et al.: A new hybrid (hydro-numerical) model of the circulatory system. Bull. Pol. Ac. Tech. **61**(4), 993–1003 (2013)
6. Sintek, M., Gdowski, M., Lindman, B., et al.: Intra-aortic balloon counterpulsation in patients with chronic heart failure and cardiogenic shock: clinical response and predictors of stabilization. J. Card. Fail. **21**(11), 868–876 (2015)
7. Santa-Cruz, R., Cohen, M., Ohman, E., et al.: Aortic counterpulsation: a review of the hemodynamic effects and indications for use. Catheter Cardiovasc. Interv. **67**(1), 68–77 (2006)
8. Kolyva, C., Biglino, G., Pepper, J.R., Khir, A.W.: A mock circulatory system with physiological distribution of terminal resistance and compliance: application for testing the intra-aortic balloon pump. Artif. Organs. **36**(3), E62-70 (2012)
9. Ferrari, G., Khir, A., Fresiello, L., et al.: Hybrid model analysis of intra-aortic balloon pump performance as a function of ventricular and circulatory parameters. Artif. Organs. **35**(9), 902–911 (2011)
10. Fresiello, L., Khir, A., Di Molfetta, A., et al.: Effects of intra-aortic balloon pump timing on baroreflex activities in a closed-loop cardiovascular hybrid model. Artif. Organs. **37**(3), 237–247 (2013)
11. Mottram, P., Haluska, B., Leano, R., et al.: Relation of arterial stiffness to diastolic dysfunction in hypertensive heart disease. Heart **91**(12), 1551–1556 (2005)
12. Herrington, D., Brown, W., Mosca, L., et al.: Relationship between arterial stiffness and subclinical aortic atherosclerosis. Circulation **110**(4), 432–437 (2004)
13. Ferrari, G., Kozarski, M., Zieliński, K., et al.: A modular computational circulatory model applicable to VAD testing and training. J. Artif. Organs. **15**(1), 32–43 (2012)
14. Shi, Y., Korakianitis, T., Bowles, C.: Numerical simulation of cardiovascular dynamics with different types of VAD assistance. J. Biomech. **40**(13), 2919–2933 (2007)
15. Downey, J.M., Kirk, E.S.: Inhibition of coronary blood flow by a vascular waterfall mechanism. Circ. Res. **36**(6), 753–760 (1975). https://doi.org/10.1161/01.res.36.6.753. PMID: 1132069

Analyzing the Sensitivity of a Procedure for Obtaining a Spherical Contact Pair to Model the Hip Joint

Joanna Rymek and Adam Ciszkiewicz(✉)

Faculty of Mechanical Engineering, Cracow University of Technology, Cracow, Poland
joanna.rymek@onet.pl, adam.ciszkiewicz@pk.edu.pl

Abstract. The geometry of the hip joint elements has been the main topic of multiple studies. Many different approaches have been proposed. In most cases, the research is focused only on one of the bones in the joint, either the femur or the pelvis. On the other hand, our research was focused on the interaction between the femur and the pelvis, with their geometry simplified into spheres. The bones then formed a simple sphere-sphere contact pair. The main aim of this study was to analyze how the different approaches to obtaining the spheres for the bones affects the obtained contact pair. In order to do it, we created fifty-six, contact pairs based on: five sets of computer tomography scans, four approaches to contact region selection and two sphere-fitting methods—with optimization and using a simplified, linearized approach. Furthermore, we also created three unique three-dimensional models from one medical set, but with different segmentation parameters. This allowed us to study and simulate the subjective aspects of medical image processing and their effect on the final biomechanical model. Our results indicated that the contact region selection had a major impact on the quality of the obtained contact pair. The best results were obtained when selecting a lunar-shaped patch on both the femur and the pelvis. Furthermore, subjective aspects of processing the medical scans also affected the results—the obtained pairs were of different geometry. In terms of the numerical approach to sphere fitting—the more general, optimizational approach resulted in better fit spheres.

Keywords: Contact pair · Bone · Sphere fitting

1 Introduction

The hip joint is an important structure in the lower leg, which plays a significant role in ambulation, while also providing stability. It is often referred to as a ball-socket joint, while its dimensions make it the largest joint of this kind in the human body. Anatomically, the joint connects the femur and a part of the pelvis called acetabulum. This connection is realized through soft tissue—cartilage—and a complex system of ligaments. Multiple muscles also play an important role in articulating and stabilizing this structure.

D. G. Pijanowska et al. (eds.), *Biocybernetics and Biomedical Engineering – Current Trends and Challenges*, Lecture Notes in Networks and Systems 293,
https://doi.org/10.1007/978-3-030-83704-4_11

The geometry of the articulating surfaces of the bones in the hip joint has been an active research topic for many years, as seen in [1–3]. Understandably, various approaches to hip contact modeling have been proposed. In the area of contact modeling of the lower limb, three major approaches can be distinguished:

- fully deformable contact elements with complex geometry using finite element method (FEM) [4–8],
- deformable contact pairs with simplified geometry using multibody system method (MBS) [9–13],
- rigid contact pairs with simplified geometry using MBS [14–17].

A significant advantage of FEM is that, using it, almost arbitrarily complex geometry can be employed for contacting surfaces. Furthermore, the elements can also express nonlinear material properties [8], which can be of importance when modeling cartilage. Nevertheless, the method requires significant computational power, which makes it difficult to use in complex problems taking advantage of optimization. On the other hand, in the multibody approach, elements with complex geometry are usually replaced with simple mechanical counterparts. For instance, a complex surface of the cartilage is can be simplified with a spherical contact pair [9].

To the best of our knowledge, no study has yet analyzed, whether the process of obtaining spherical pairs for the hip joint can be affected by subjective aspects of three-dimensional modeling and numerical approaches for shape fitting. Therefore, the aim of this study was to analyze the sensitivity of a procedure for obtaining a spherical contact pair to model the hip joint. In order to do so, we obtained five CT datasets (available under the Creative-Commons license), created 3D surface models of the femur/pelvis (one model for each dataset and two additional models on the first datasets, but with a slightly different procedure). From the seven surface models we cut 28 different fragments of the meshes, which could be in potential contact during hip motion, and fit spheres to them with two different methods. This resulted in 56 unique spherical contact pairs, which were then rated and analyzed with regards to their usability in hip models and the overall sensitivity of the procedure. The details regarding our approach and the obtained results were described in the following paragraphs.

2 Method

As mentioned in the introduction our aim was to analyze the sensitivity of the procedure for obtaining a contact pair to model the hip joint. In this section we presented the details regarding the procedure for obtaining the actual contact pairs as well as our method to assess the results.

2.1 Obtaining the Meshes of the Bones

In this study we employed *3D Slicer* to obtain three-dimensional meshes representing the femur and the pelvis based on actual CT medical scans. The procedure was typical and consisted of thresholding, with manually set thresholds, manual improvement of the selection and marching cubes to turn the segmented image into the surface mesh. The obtained geometry was then corrected using the functions: *joint smoothing* and *mediane*, available in *3D Slicer*—see Fig. 1 for a sample mesh of the femur.

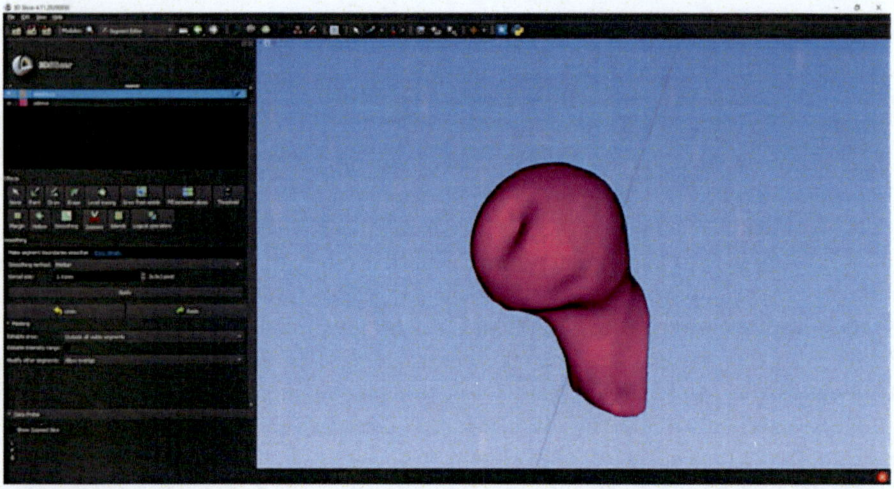

Fig. 1 One of the seven obtained surface meshes of the femur head, as seen in 3D Slicer

The aforementioned procedure was repeated for five separate CT datasets of the lower limb available under Creative Commons By Attribution license at embodi3d.com and downloaded at 08.01.2021. The details regarding the datasets were presented in Table 1.

Table 1 The datasets used in this study

Author (username)	*Filename*	*Date*	*License*
Pam Waterman	*SpineTest2 1.0.0*	*09.11.2020*	*CC—Attribution*
Carmenalb1998	*Abdomen*	*14.11.2020*	*CC—Attribution*
Aldo Machuca	*TEST 1.0.0*	*07.01.2021*	*CC—Attribution*
felipefavadelima	*TK full abdomen 1.0.0*	*21.10.2020*	*CC—Attribution*
Rubenz	*Plevis Tumor*	*19.10.2020*	*CC—Attribution*

Additionally, the procedure was repeated three times for the first dataset with altered parameters of thresholding and further processing of the surface. This allowed us to not only test the approach under patient-specific conditions, but also take into account subjective decisions made during the modeling process. In total, we obtained seven pairs of meshes of the femur and the pelvis.

2.2 Cutting the Meshes and Assembling the Contact Pairs

The obtained meshes of the bones were imported into *Meshmixer*, in which patches of them were selected and cut. These cuts represented the surfaces, based on which the contact pairs could be obtained. In order to verify how sensitive the approach was to patch selection, we used four, different cutting schemes for each femur-pelvis mesh pair. These schemes were presented visually in Fig. 2. The lunar shape was after [18].

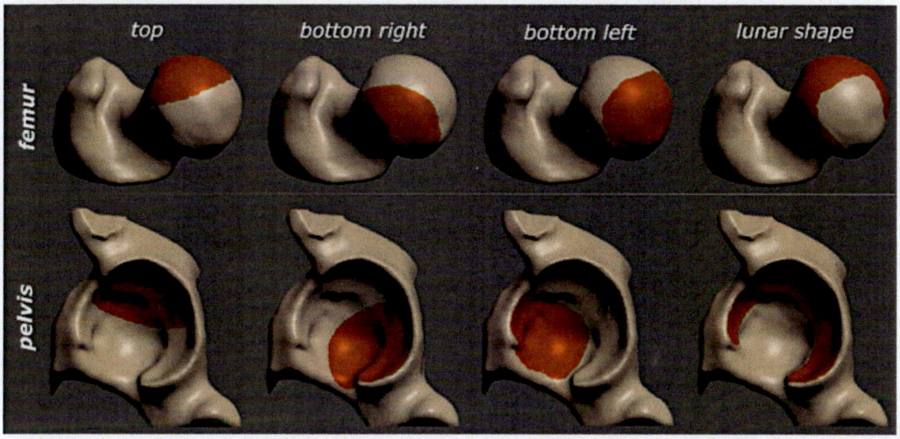

Fig. 2 The mesh selection method employed in this study

After the selection procedure, each mesh was fitted to a sphere with an optimizational approach in *Python* using a *least_sq* function in *scipy* library [19]. The code was an extension of the procedure for fitting circles available at [20]. To verify the results, we also applied a simpler, linearized method for sphere fitting [20]—referred to as *Approximation* in the later part of the paper. The obtained spheres and their corresponding meshes were then visualized in pairs—femur/pelvis—using *mayavi*, also in *Python*. A sample contact pair was presented in Fig. 3. Considering four patches for each mesh, we obtained 56 (7 models * 4 selections * 2 methods) unique contact pairs.

Note that we used the bone profiles based on CT scans to determine the contact pairs in the hip joint, which is a common approach in biomechanical modeling, as seen in [21]. In the joint, the actual contact is realized through the cartilage. The cartilage thickness could be taken into account after obtaining the contact pairs from the bone profiles, for instance, based on experimental values [22, 23]. This simplified approach was sufficient to fulfill the main aims of the paper.

2.3 Assessing the Quality of a Contact Pair

Our main goal was to assess how sensitive the procedure for obtaining contact pairs (see Sects. 2.1 and 2.2) was to both the patient-specific geometrical features and the subjective aspects of the 3D modeling process. In order to do so, we performed two

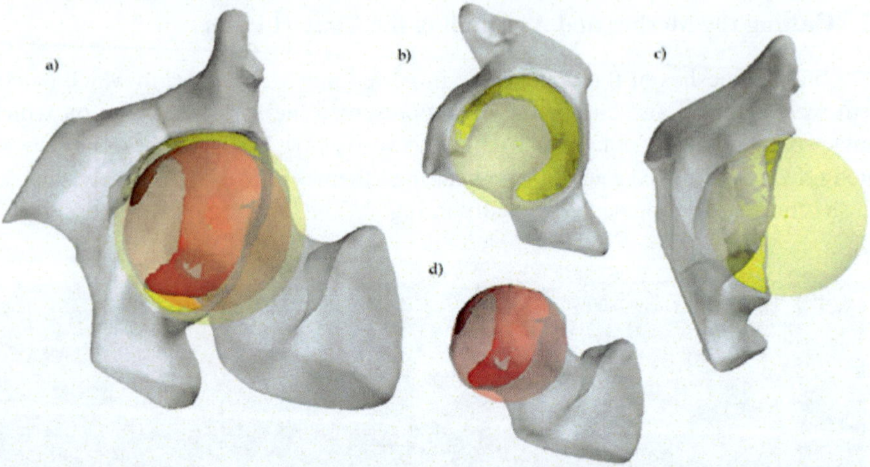

Fig. 3 A sample contact pair between the femur and the pelvis obtained using the aforementioned procedure, where: **a** the full contact pair, **b** and **c** the sphere fitted to the pelvis and **d** the sphere fitted to the femur

different analyses. Firstly, all of the obtained contact pairs were rated based on their validity. We distinguished three cases:

- a valid contact pair—1.0 point—the sphere for the femur was smaller than that of the pelvis, and the spheres did not intersect
- a partially-valid contact pair—0.5 point—the femur sphere was smaller than the pelvis sphere, but the spheres intersected, albeit far from the initial contact patches,
- an invalid contact pair—0.0 points—the femur sphere was larger than the pelvis sphere—this case represented a pair, which was in conflict with the anatomy of the joint.

The three cases were visualized in Fig. 4.

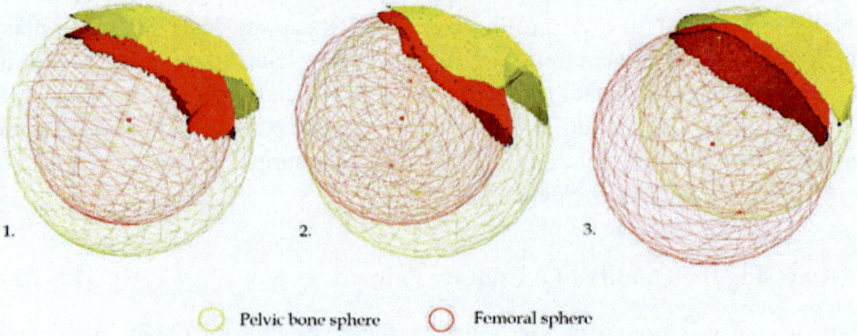

Fig. 4 A visual assessment of the obtained contact pairs

This allowed us to check, how selecting the mesh region for sphere fitting affected the results.

In the second case, we used the three variant of the first model, as mentioned in the Sect. 2.1 and compared the radii and the standard deviation of the spheres for both the femur and the pelvis in these three submodels for all the four mesh selection approaches. These results were used to assess how sensitive the procedure was to subjective aspects of 3D modelling for use in biomechanical models.

3 Results and Discussion

3.1 The Visual Assessment

As mentioned before, our main assessment procedure was based on the visual representation of a contact pair. It allowed us to ascertain whether a pair could be implemented in a physical model of the hip joint without any further modifications to its geometry. This assessment was performed on each of the 56 unique contact pairs based on 4 approaches to contact patch selection, as specified in the Sect. 2.1. The obtained results, presented in Figs. 5 and 6, showed that regardless of the method used for sphere fitting, the best contact pairs were obtained using the lunar-shaped cut of the femur and the pelvis meshes. In all of the tested scenarios, this cutting approach resulted in viable contact pairs, with anatomically-correct sphere placement and size proportions.

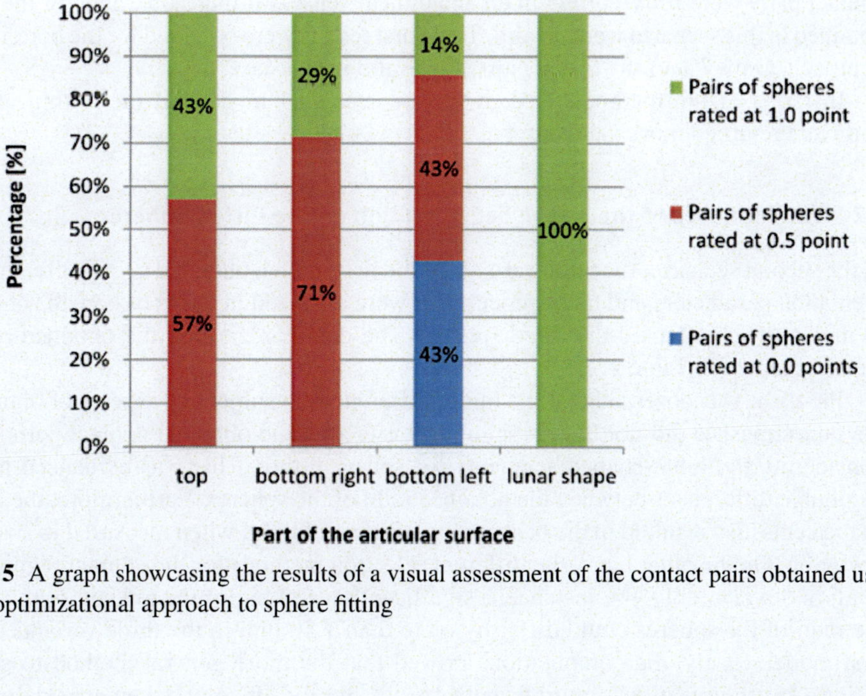

Fig. 5 A graph showcasing the results of a visual assessment of the contact pairs obtained using the optimizational approach to sphere fitting

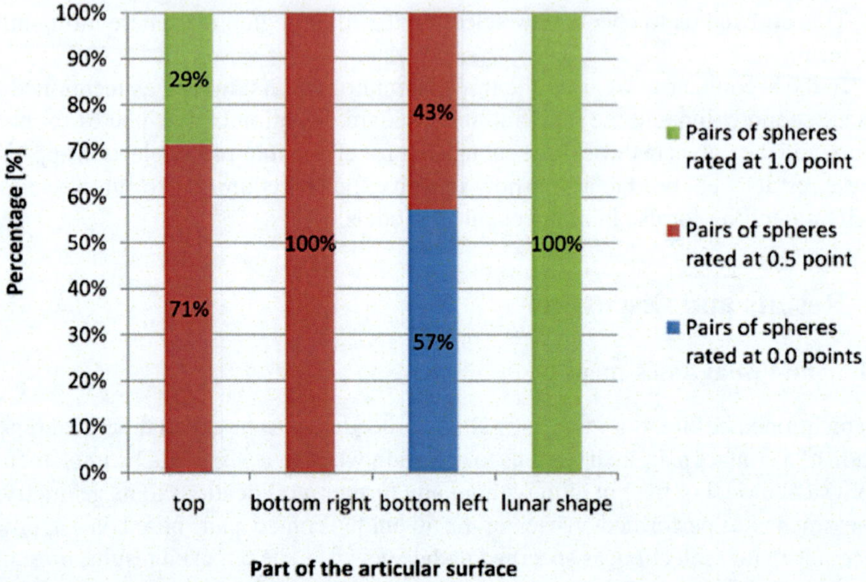

Fig. 6 A graph showcasing the results of a visual assessment of the contact pairs obtained using the approximational approach to sphere fitting

The second best approach was the top cut. In this case, on average 36% of the contact pairs were fully-correct in an anatomical sense. Furthermore, none of the pairs obtained in this scenario were invalid. The worst results were obtained for the bottom-left approach. Only 7% of the tested pairs were valid in this case.

In terms of the methods used, it can be seen that in general the optimizational approach returned more valid cases.

3.2 The Influence of the Image Segmentation on the Fitted Spheres

In the second scenario, the three variants of the first model, obtained using different segmentation parameters and mesh processing, were analyzed in terms of how the changes to the approach affected the fitted spheres. The details regarding the obtained results were presented in Table 2.

Firstly, it was observed that the minor differences in subjective aspects of obtaining the bones meshes did not have a significant effect on the obtained contact pairs when considering the lunar-shaped segments for sphere fitting. This was evidenced by the negligible differences between the obtained radii of the spheres. Furthermore, the lunar-shaped cuts also resulted in the best quality of sphere fitting, when measured in standard deviation. On the other hand, the different selection methods for the contact regions did not turn out favorably. For instance, a significant sensitivity for the top cut. In this case, the radii of the spheres could differ by more than 1.50 mm in the three variants of the first model. Lastly, the computations proved that the more general method to sphere fitting—by optimization—returned better results for the hip joint, when considering the standard deviation of the fits.

Table 2 The details regarding the spheres fitted in the three variants of the first model, where *M*, *O*, *A*, *F*, *P*, *r*, σ stand for *Model, Optimizational method*, *Approximation method*, *Femur*, *Pelvis*, *radius*, *standard deviation*

M			*Top*		*Bottom right*		*Bottom left*		*Lunar shape*	
			r [mm]	σ [mm]	*r* [mm]	σ [mm]	*r* [mm]	σ [mm]	*r* [mm]	σ [mm]
1.1	***O***	***F***	27.08	1.22	25.90	1.44	24.93	1.55	26.38	0.44
		P	22.72	0.76	23.63	0.74	25.99	0.73	24.10	0.38
	A	***F***	25.79	1.25	24.62	1.48	23.87	1.59	26.37	0.44
		P	22.61	0.76	23.40	0.75	25.36	0.74	24.10	0.38
1.2	***O***	***F***	25.62	1.03	25.60	1.43	24.61	1.74	26.24	0.41
		P	23.41	0.70	24.47	0.68	25.80	0.61	24.45	0.33
	A	***F***	24.80	1.05	24.69	1.46	23.76	1.77	26.24	0.41
		P	23.21	0.71	24.24	0.68	25.40	0.61	24.44	0.33
1.3	***O***	***F***	26.76	1.44	25.95	1.37	23.67	1.42	26.10	0.41
		P	24.00	0.78	24.05	0.87	26.27	0.77	24.46	0.84
	A	***F***	25.31	1.49	24.71	1.41	22.79	1.45	26.09	0.41
		P	23.74	0.79	23.64	0.87	25.37	0.79	24.46	0.84

4 Conclusion

This research was focused on the interaction between the femur and the pelvis, with their geometry simplified into spheres. We assumed that the bones formed a simple sphere-sphere contact pair. The main aim of this study was to analyze how the different approaches to obtaining the spheres for the bones affect the contact pairs in terms of their validity and sphere-fit quality. In order to do so, we created fifty-six, unique contact pairs and analyzed their geometrical features. Furthermore, we also created three models from one medical dataset, but with different segmentation parameters, which allowed us to study and simulate the subjective aspects of medical image processing and their effect on the obtained contact pairs.

Our findings indicated that the contact region selection had a major impact on the quality of the obtained contact pair. The best results were obtained when selecting.

a lunar-shaped patch on both the femur and the pelvis. Furthermore, subjective aspects of processing of the medical scans also affected the results, albeit only in certain approaches to contact region selection. Again, the lunar-shaped patches proved to be the most stable. In terms of the numerical approach to sphere fitting—the more general, optimizational approach resulted in better fit spheres.

References

1. Cerveri, P., Marchente, M., Chemello, C., Confalonieri, N., Manzotti, A., Baroni, G.: Advanced computational framework for the automatic analysis of the acetabular morphology from the pelvic bone surface for hip arthroplasty applications. Ann. Biomed. Eng. **39**, 2791–2806 (2011). https://doi.org/10.1007/s10439-011-0375-5
2. Cerveri, P., Manzotti, A., Baroni, G.: Patient-specific acetabular shape modelling: comparison among sphere, ellipsoid and conchoid parameterisations (2014). http://dx.doi.org/10.1080/10255842.2012.702765
3. Lopes, D.S., Neptune, R.R., Gonçalves, A.A., Ambrósio, J.A., Silva, M.T.: Shape analysis of the femoral head: a comparative study between spherical, (super)ellipsoidal, and (super)ovoidal shapes. J. Biomech. Eng. **137**, 1–8 (2015). https://doi.org/10.1115/1.4031650
4. Zhang, H., Lv, M.L., Yang, J., Niu, W., Cheung, J.C.W., Sun, W., Wong, D.W.C., Ni, M.: Computational modelling of foot orthosis for midfoot arthritis: a Taguchi approach for design optimization. Acta Bioeng. Biomech. **22**, 1–17 (2020). https://doi.org/10.37190/ABB-01694-2020-03
5. Bashiri, A., Sallam, H.E.M., Abd-Elhady, A.A.: Progressive failure analysis of a hip joint based on extended finite element method. Eng. Fail. Anal. **117**, 104829 (2020). https://doi.org/10.1016/j.engfailanal.2020.104829
6. Li, M., Venäläinen, M.S., Chandra, S.S., Patel, R., Fripp, J., Engstrom, C., Korhonen, R.K., Töyräs, J., Crozier, S.: Discrete element and finite element methods provide similar estimations for hip joint contact mechanics during walking gait. J. Biomech. **115**, 110163 (2021). https://doi.org/10.1016/j.jbiomech.2020.110163
7. Ma, H., Mao, Z., Li, G., Yan, L., Mo, F.: Could an isolated human body lower limb model predict leg biomechanical response of Chinese pedestrians in vehicle collisions. Acta Bioeng. Biomech. **22**, 1–28 (2020). https://doi.org/10.37190/ABB-01630-2020-04
8. Li, J.: Development and validation of a finite-element musculoskeletal model incorporating a deformable contact model of the hip joint during gait. J. Mech. Behav. Biomed. Mater. **113**, 104136 (2021). https://doi.org/10.1016/j.jmbbm.2020.104136
9. Askari, E., Flores, P., Dabirrahmani, D., Appleyard, R.: Nonlinear vibration and dynamics of ceramic on ceramic artificial hip joints: a spatial multibody modelling. Nonlinear Dyn. **76**, 1365–1377 (2014). https://doi.org/10.1007/s11071-013-1215-y
10. Borucka, A., Ciszkiewicz, A.: A planar model of an ankle joint with optimized material parameters and Hertzian contact pairs. Materials (Basel) **12**, 2621 (2019)
11. Ciszkiewicz, A.: Analyzing uncertainty of an ankle joint model with genetic algorithm. Materials (Basel) **13**, 1175 (2020). https://doi.org/10.3390/ma13051175
12. Machado, M., Flores, P., Claro, J.C.P., Ambrósio, J., Silva, M., Completo, A., Lankarani, H.M.: Development of a planar multibody model of the human knee joint. Nonlinear Dyn. **60**, 459–478 (2009). https://doi.org/10.1007/s11071-009-9608-7
13. Machado, M., Flores, P., Ambrosio, J., Completo, A.: Influence of the contact model on the dynamic response of the human knee joint. Proc. Inst. Mech. Eng. Part K J. Multi-body Dyn. **225**, 344–358 (2011). https://doi.org/10.1177/1464419311413988
14. Apkarian, J., Naumann, S., Cairns, B.: A three-dimensional kinematic and dynamic model of the lower limb. J. Biomech. **22**, 143–155 (1989)
15. Parenti-Castelli, V., Leardini, A., Gregorio, R., O'Connor, J.: On the modeling of passive motion of the human knee joint by means of equivalent planar and spatial parallel mechanisms. Aut. Robot. **16**, 219–232 (2004). https://doi.org/10.1023/B:AURO.0000016867.17664.b1
16. Ottoboni, A., Parenti-Castelli, V., Sancisi, N., Belvedere, C., Leardini, A.: Articular surface approximation in equivalent spatial parallel mechanism models of the human knee joint: an experiment-based assessment. Proc. Inst. Mech. Eng. H. **224**, 1121–1132 (2010)

17. Michnik, R., Zadoń, H., Nowakowska-Lipiec, K., Jochymczyk-Woźniak, K., Myśliwiec, A., Mitas, A.W.: The effect of the pelvis position in the sagittal plane on loads in the human musculoskeletal system. Acta Bioeng. Biomech. **22**, 1–18 (2020). https://doi.org/10.37190/ABB-01606-2020-02
18. Hammond, A.S., Plavcan, J.M., Ward, C.V.: A validated method for modeling anthropoid hip abduction in silico. Am. J. Phys. Anthropol. **160**, 529–548 (2016). https://doi.org/10.1002/ajpa.22990
19. van der Walt, S., Colbert, S.C., Varoquaux, G.: The NumPy array: a structure for efficient numerical computation. Comput. Sci. Eng. **13**, 22–30 (2011). https://doi.org/10.1109/MCSE.2011.37
20. Various: Least squares circle. https://scipy-cookbook.readthedocs.io/items/Least_Squares_Circle.html
21. Wei, F., Hunley, S.C., Powell, J.W., Haut, R.C.: Development and validation of a computational model to study the effect of foot constraint on ankle injury due to external rotation. Ann. Biomed. Eng. **39**, 756–765 (2011). https://doi.org/10.1007/s10439-010-0234-9
22. Shepherd, D.E.T., Seedhom, B.B.: Thickness of human articular cartilage in joints of the lower limb. Ann. Rheum. Dis. **58**, 27–34 (1999). https://doi.org/10.1136/ard.58.1.27
23. Koo, S., Gold, G.E., Andriacchi, T.P.: Considerations in measuring cartilage thickness using MRI: factors influencing reproducibility and accuracy. Osteoarthr. Cartil. **13**, 782–789 (2005). https://doi.org/10.1016/j.joca.2005.04.013

Fourier Transform Layer for Fast Foreground Segmentation in Samples' Images of Tissue Biopsies

Jakub Zak[1(✉)], Krzysztof Siemion[1,2], Lukasz Roszkowiak[1], and Anna Korzynska[1]

[1] Laboratory of Processing Systems of Microscopic Image Information, Nalecz Institute of Biocybernetics and Biomedical Engineering Polish Academy of Sciences, Ks. Trojdena 4, 02-109 Warsaw, Poland
jzak@ibib.waw.pl

[2] Medical Pathomorphology Department, Medical University of Bialystok, Bialystok, Poland

Abstract. Sjőgren's Syndrome is a systemic disease, presenting itself in a spectrum of symptoms and signs throughout the body. Its diagnosis can include the examination of minor salivary gland biopsies. The image processing of digital images, specifically large-scale Whole Slide Images can aid in diagnosis, but unfortunately requires a lot of processing power and time for automatic analysis. This paper presents a new method for fast foreground (tissue) segmentation in fragments of Whole Slide Images containing biopsies of minor salivary glands taken from patients with diagnosed Sjőgren's Syndrome. The method involves training a neural network with a novel type of layer, which calculates Fourier transform of data within the network. The neural network achieved accuracy of 89% on a test dataset.

Keywords: Neural networks · Fourier Transform · Image classification · WSI

1 Introduction

In recent years, interest in neural networks (NN) has increased. One area, where a specific type of NN—Convolutional Neural Networks (CNNs)—performs very well, is image processing. During recent years, CNNs have been successfully applied to various image processing tasks [1,2]. The image processing research is becoming dominated by CNN research.

One of the challenges is the limit of the data (image) size of a single sample. It is interlinked with training time: the larger the images, the longer it takes to process a sample. This can also quickly increase memory (RAM or GPU) requirements. Due to those limitations, networks are trained on very small images, in comparison to contemporary commercial standards of typical image acquisition tools. Images in datasets are often resized (downsampled), resulting in lost fine details.

D. G. Pijanowska et al. (eds.), *Biocybernetics and Biomedical Engineering – Current Trends and Challenges*, Lecture Notes in Networks and Systems 293,
https://doi.org/10.1007/978-3-030-83704-4_12

1.1 Sjőgren's Syndrome

Sjőgren's Syndrome (SS) is a systemic disease. A spectrum of symptoms and signs [3] throughout the entire body can reveal it. The diagnosis is based on an assessment of symptoms, signs, clinical tests, the concentration of antibodies in the patient's blood and histopathological examination of minor salivary glands [4]. Examination of salivary glands may be helpful in a process of diagnosis [5,6]. Pathologists use a measure called focus score [7] to semi-quantitatively evaluate the changes caused by focal inflammation of the salivary gland in SS. However, human estimation is inherently subjective, which often leads to diverging diagnosis, because of inter and intra-observer variability during hispotathological slide assessment.

1.2 Aim

The aim of this article is to present the results of our approach to fast pre-selection of WIS parts by training a neural network with FT. The proposed Fourier Neural Network (FNN) contained only implementation of Fourier Transform Layer (FTL) (described in Sect. 2.2), Flatten for 1D reshaping of data and Fully Connected Layer for classification at the end. A simple schematic is presented in Fig. 1.

1.3 Related Work

There is a lot of research on tissue segmentation in WSI. The most recent [8–12], mainly focus on using one of the well established CNN architectures to perform the segmentation, with some exceptions [13]. The CNNs achieve very high accuracy. However, the researchers do not mention the speed of the processing, which is important for the users—pathologists. [8,9,11] present CNNs trained on a dataset of tiles (also called patches), whereas authors of [10] and [12] incorporate U-net with different backbones to segment the tissue.

The notion of combining FT and NN application for data analysis can be traced as far back as 1999 [14], where the FT of an electrocardiogram (ECG or EKG) was used as a preprocessing step before training a simple NN on the data. Most of the research is focused either on improving the speed of convolutions by FFT [15–17] or preprocessing the data using FFT [18–22].

In the literature, research using FT as a preprocessing tool is split mainly into two areas: typical FT [14,20,23] or Short Time Fourier Transform [18,19,21,24] (STFT). Advantages of the first one have already been explained [14,18–20,20–23] and can be surmised as fast filtering or transformation of data representation.

2 Materials and Methods

2.1 Datasets Used During Training

In the presented investigation minor salivary gland biopsy slides are stained with hematoxylin and eosin (H&E), which causes lymphocytes' nuclei and other cellular nuclei to present a blue tint. The tint is caused by the hematoxylin (in nuclei) absorbing light.

The image dataset used in this investigation contained the same WSI as our previous investigation [25], but it was modified for a different aim. The modifications were: different labels. Images used in this investigation were converted to grayscale, because FFT does not discern between colors. This operation also speeds up calculations.

Table 1 presents general highlights of differences in data and processing between investigations. The classes in our previous study were marking the presence of chronic inflammatory infiltrate. As we have explained in Sect. 1.2, we decided to add a preliminary stage of analysis by excluding the areas of background from the tissue sections. For that, we have prepared an adjusted dataset, using a lower number of WSI with different annotations made by pathologists.

Table 1 Highlights. Note: No. of WSI means the number of WSI fragments

Feature	Previous research [25]	This article
Annotation by	Pathologist	Pathologist
No. of WSI (train/test)	107 (103/4)	19 (14/5)
Tiling (size/step)	256/64	224/56
Color space	R,G,B	Grayscale
Approx. no. of tiles	170 000	1 430 000
No. of classes	2	2
Class meaning	0—no inflammation	0—background
	1—inflammation	1—foreground (tissue)

The large discrepancy between the number of training/test tiles (see: approx. no. of tiles Table 1) is caused by a greater number of tiles containing images of salivary gland parenchyma, compared to the number of tiles containing images of inflammatory infiltration. Previously, to avoid large imbalances between classes, only some of the tiles containing overrepresented class (0) were used [25]. In this investigation, training tiles were not augmented regardless of imbalance in classes—2.4:1 ratio (0 s:1 s).

2.2 Fourier Transform Layer

FTL is a neural network layer, calculating two-dimentional FFT on the input data. After FFT, the real part of the transform is multiplied by the weights of the FTL. The imaginary part is not multiplied by any weights in the current implementation. The number of weight parameters in the layer is X^2, where X is both axes dimension of the input image. This number is the same as the size (number of pixels) of the image.

FNN operates in the following way. An image is an Input. FTL performs 2DFFT, multiplies the result by FTL weights and returns an absolute of real and imaginary parts. Then, the data is reshaped into 1D matrix (Flatten). Finally, a Fully Connected (or Dense) layer is used for classification. Figure 1 presents the scheme.

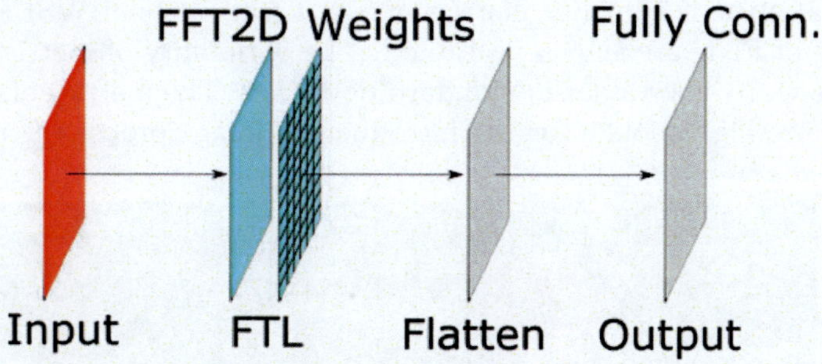

Fig. 1 A simple schematic of a FNN with one FTL. This kind of architecture is commonly used for classification tasks (Fully Connected layer at the top of NN)

2.3 Evaluation of Results

By superimposing the inflammatory infiltration map resulting from the FNN onto a corresponding ground truth mask (pathologists' annotation), true positive (TP), true negative (TN), false positive (FP), and false negative (FN) pixels were calculated. Using TN, TP, FP, and FN, the metrics such as F-score, accuracy, specificity, precision, and recall were calculated to evaluate the results of classification by FNN. F–score, accuracy and recall were already defined in [25].

Additional metrics are defined as:

- Precision:

$$P = \frac{TP}{TP + FP} \tag{1}$$

- Specificity:

$$S = \frac{TN}{TN + FP} \tag{2}$$

2.4 Hardware and Software Used

The tests were conducted both on GPU and CPU. GPU—NVIDIA GeForce GTX 1060 with 6 GB memory. CPU—Intel Core i5-6600K @ 3.50 GHz with 64 GB memory. The code was written in Python 3.6+ using Keras (v2.2.4) and Tensorflow (v1.12) for neural networks.

3 Results

FNN plateaued during training at around 83% accuracy.

Figure 2 presents the best result map from classified image tiles, in comparison to the ground truth. The boundaries between tissue and background are poorly separated, there are also a few FP areas on the map.

The statistics of 5 quality metrics of 5 test fragments of WSI after 2 epochs of training of FNN are shown in Table 2. The variability of metrics is relatively low, as shown by low values in standard deviation. The metrics of accuracy and specificity are high enough for the introductory image processing run.

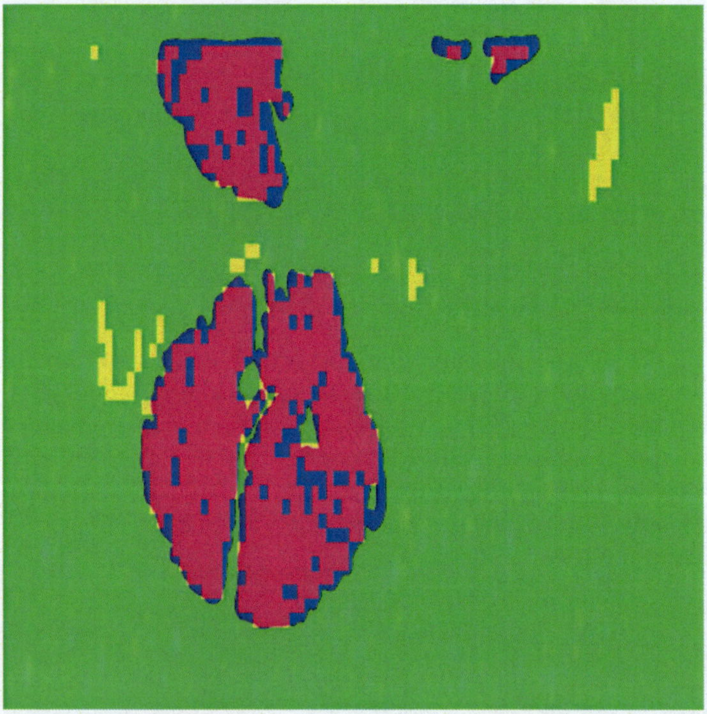

Fig. 2 A map of TP (magenta), TN (green), FP (yellow) and FN (blue) on a test fragment WSI

Table 2 Statistics of performance metrics on a testing set of 5 fragments of WSI

Measure	Mean	Standard deviation	Median
F–Score	0.72	0.10	0.69
Accuracy	0.89	0.04	0.87
Precision	0.81	0.11	0.84
Recall	0.66	0.14	0.64
Specificity	0.96	0.03	0.97

4 Discussion

Figure 2 shows the results of using FNN on large tiled biomedical images, like WSI, which are promising. FNN can greatly accelerate the selection of tiles containing images of tissues, thus reducing the time of processing by CNNs trained to detect inflammation [25].

Our experience shows that FTL significantly improves training and inference times, compared to typical CNNs, when details in images are lost by use of resizing processing. Comprehensive studies on the gain in speed are planned.

However, the proposed solution requires fine-tuning before incorporation into the foreground extraction pipeline. Possible solutions include: increasing the number of WSI fragments, using augmentation methods, and using other tissue datasets prepared by various laboratories to provide the higher variability. This should increase the accuracy reached during the process of FNN training.

5 Conclusions and Future Work

To sum up, a novel method of foreground segmentation in WSI by the use of a NN containing the FTL was presented.

This investigation shows FNN training on the medical image dataset achieved good results. However, we would like to modify the training pipeline to focus on decreasing false negative rate, thus improving Recall, which could increase the FP values. However, FNN was prepared to be used as a fast pre-selection tool, so lower values of FP are acceptable. This way, no tile containing the histological image would be left unprocessed by CNNs. The FNN is not limited only to images: the code we presented can be easily adjusted to one-dimentional data. Therefore we would like to invite the scientific community to collaborate with us on research regarding the possible applications of the FTL. We focus on processing medical images, but would gladly welcome co-operation on biomedical data of any suitable kind.

References

1. Dhruv, P., Naskar, S.: Image classification using convolutional neural network (CNN) and recurrent neural network (RNN): a review. In: Machine Learning and Information Processing, pp. 367–381. Springer, Singapore (2020). https://doi.org/10.1007/978-981-15-1884-3_34
2. Rawat, W., Wang, Z.:"Deep convolutional neural networks for image classification: a comprehensive review. Neural Comput. **29**(9), 2352–2449 (2017). https://doi.org/10.1162/neco_a_00990
3. Hellquist, H., Skalova, A.: Histopathology of the Salivary Glands. Springer, Heidelberg (2014). https://doi.org/10.1007/978-3-540-46915-5
4. Varela-Centelles, P., Seoane-Romero, J., Sanchez-Sanchez, M., Gonzalez-Mosquera, A., Diz-Dios, P., Seoane, J.: Minor salivary gland biopsy in Sjogren's syndrome: a review and introduction of a new tool to ease the procedure, Medicina Oral Patología Oral y Cirugia Bucal e20–e23 (2014). https://doi.org/10.4317/medoral.19131

5. Shiboski, C.H., et al.: 2016 American college of rheumatology/European league against rheumatism classification criteria for primary Sjögren's syndrome: a consensus and data-driven methodology involving three international patient cohorts. Arthritis Rheumatol. **69**(1), 35–45 (2016). https://doi.org/10.1002/art.39859
6. Costa, S., et al.: B-cell and t-cell quantification in minor salivary glands in primary Sjögren's syndrome: development and validation of a pixel-based digital procedure. Arthritis Res. Therapy **18**(1). https://doi.org/10.1186/s13075-016-0924-2
7. Waterhouse, J.P.: Focal adenitis in salivary and lacrimal glands. Proc. Roy. Soc. Med. **56**, 911–918 (1963)
8. Muñoz-Aguirre, M., Ntasis, V.F., Rojas, S., Guigó, R.: PyHIST: a histological image segmentation tool. PLoS Comput. Biol. **16**(10), e1008349 (2020). https://doi.org/10.1371/journal.pcbi.1008349
9. Bándi, P., Balkenhol, M., van Ginneken, B., van der Laak, J., Litjens, G.: Resolution-agnostic tissue segmentation in whole-slide histopathology images with convolutional neural networks. PeerJ 7, e8242 (2019). https://doi.org/10.7717/peerj.8242
10. Feng, R., Liu, X., Chen, J., Chen, D.Z., Gao, H., Wu, J.: A deep learning approach for colonoscopy pathology WSI analysis: accurate segmentation and classification. IEEE J. Biomed. Health Inform. 1 (2020). https://doi.org/10.1109/jbhi.2020.3040269
11. Chan, L., Hosseini, M.S., Rowsell, C., K. Plataniotis, N., Damaskinos, S.: HistoSegNet: semantic segmentation of histological tissue type in whole slide images. In: Proceedings of the IEEE/CVF International Conference on Computer Vision (ICCV) (2019)
12. Oskal, K.R.J., Risdal, M., Janssen, E.A.M., Undersrud, E.S., Gulsrud, T.O.: A U-Net based approach to epidermal tissue segmentation in whole slide histopathological images. SN Appl. Sci. **1**(7). https://doi.org/10.1007/s42452-019-0694-y
13. Kleczek, P., Jaworek-Korjakowska, J., Gorgon, M.: A novel method for tissue segmentation in high-resolution H&E-stained histopathological whole-slide images. Comput. Med. Imaging Graph. **79**, 101686 (2020). https://doi.org/10.1016/j.compmedimag.2019.101686
14. Minami, K., Nakajima, H., Toyoshima, T.: Real-time discrimination of ventricular tachyarrhythmia with fourier-transform neural network. IEEE Trans. Biomed. Eng. **46**(2), 179–185 (1999). https://doi.org/10.1109/10.740880
15. Lin, J., Yao, Y.: A fast algorithm for convolutional neural networks using tile-based fast fourier transforms. Neural Process. Lett. **50**(2), 1951–1967 (2019). https://doi.org/10.1007/s11063-019-09981-z
16. Ayat, S.O., Khalil-Hani, M., Ab Rahman, A.A.H., Abdellatef, H.: Spectral-based convolutional neural network without multiple spatial-frequency domain switchings. Neurocomputing **364**, 152–167 (2019). https://doi.org/10.1016/j.neucom.2019.06.094
17. Vasilache, N., Johnson, J., Mathieu, M., Chintala, S., Piantino, S., Lecun, Y.: Fast convolutional nets with fbfft: A GPU performance evaluation (2014)
18. Zhang, Y., Xing, K., Bai, R., Sun, D., Meng, Z.: An enhanced convolutional neural network for bearing fault diagnosis based on time–frequency image. Measurement **157**, 107667 (2020). https://doi.org/10.1016/j.measurement.2020.107667
19. Ganapathy, N., Veeranki, Y.R., Swaminathan, R.: Convolutional neural network based emotion classification using electrodermal activity signals and time-frequency features. Expert Syst. Appl. 159, 113571(2020). https://doi.org/10.1016/j.eswa.2020.113571

20. Boldsaikhan, E., Corwin, E.M., Logar, A.M., Arbegast, W.J.: The use of neural network and discrete fourier transform for real-time evaluation of friction stir welding, Appl. Soft Comput. **11**(8), 4839–4846 (2011). https://doi.org/10.1016/j.asoc.2011.06.017
21. Daldal, N., Cömert, Z., Polat, K.: Automatic determination of digital modulation types with different noises using convolutional neural network based on time–frequency information. Appl. Soft Comput. **86**, 105834 (2020). https://doi.org/10.1016/j.asoc.2019.105834
22. Zeng, Y., Zhang, M., Han, F., Gong, Y., Zhang, SJ.: pectrum analysis and convolutional neural network for automatic modulation recognition, IEEE Wirel. Commun. Lett. **8**(3), 929–932 (2019). https://doi.org/10.1109/lwc.2019.2900247
23. Gothwal, H., Kedawat, S., Kumar, R.: Cardiac arrhythmias detection in an ECG beat signal using fast fourier transform and artificial neural network. J. Biomed. Sci. Eng. **04**(04), 289–296 (2011). https://doi.org/10.4236/jbise.2011.44039
24. Yao, S., et al.: STFNets: learning sensing signals from the time-frequency perspective with short-time fourier neural networks. In: The World Wide Web Conference on - WWW19. ACM Press (2019). https://doi.org/10.1145/3308558.3313426
25. Korzynska, A., Zak, J., Siemion, K., Roszkowiak, L., Pijanowska, D.: CNN support to diagnostics in sjögren's syndrome. In: Korbicz, J., Maniewski, R., Patan, K., Kowal, M. (eds.) Advances in Intelligent Systems and Computing, pp. 72–81. Springer, Cham (2019). https://doi.org/10.1007/978-3-030-29885-2_7

Corrosion Properties of Double-Walled TiO_2 Nanotubes Measured in 0.9% NaCl - Preliminary Results

Aleksandra Jędrzejewska[1,2](✉)

[1] Science and Technology Park of University of Zielona Góra, Syrkiewicza 6 Street, 66-002 Zielona Góra, Poland
a.jedrzejewska@pnt.uz.zgora.pl

[2] Department of Biomedical Engineering, Faculty of Mechanical Engineering, University of Zielona Góra, Licealna 9 Street, 65-417 Zielona Góra, Poland

Abstract. Titanium and its alloys have good corrosion resistance and biocompatibility and are often used in biomedical applications. The corrosion resistance of these materials can be improved, thereby increasing their usability and durability. The aim of this study was to improve the corrosive properties of titanium by developing a coating of double-walled titanium dioxide nanotubes (dTNT) on its surface. The dTNT layers were prepared by electrochemical anodization of titanium foils in 85% ethylene glycol solution with 0.65% wt. NH_4F and 7.5% H_3PO_4 in time 4 h using potentials: 10, 30 and 60 V. Based on the electrochemical and scanning electron microscopy studies, it was found that dTNT layer improves corrosion resistance, and the best protection was confirmed for dTNT formed at 60 V. The dTNT layer due to its good anti-corrosion, protective properties and stability can expand the applicability and extend durability of titanium in aircraft structures, medicine or chemical industries.

Keywords: Titanium dioxide (TiO_2) · Double-walled titanium dioxide nanotubes (dTNT) · Corrosion

1 Introduction

Engineering material for corrosion total resistance in any environment has not been prepared yet. Therefore, the corrosion problem is still valid and researches of the modification and preparation of new materials or coatings are desirable.

Titanium and its alloys are a special group of materials with very high electrochemical corrosion resistance. The passivation increases the corrosion resistance ability of titanium. Titanium and its alloys are being used mainly in biomedical [1], chemical processing [2], aeronautics [3], architecture [4], seawater service and arms industry [5] owing to their properties like: lightweight, good formability and high strength [6]. Titanium and its alloys are susceptible to general and local corrosion, especially in solutions containing significant amounts of fluoride

D. G. Pijanowska et al. (eds.), *Biocybernetics and Biomedical Engineering – Current Trends and Challenges*, Lecture Notes in Networks and Systems 293,
https://doi.org/10.1007/978-3-030-83704-4_13

ions, acid chemicals, strong oxidizing agents like dry chlorine, concentrated oxygen, and hot caustic solutions [7]. Corrosive wear increases operating costs and current corrosion protection methods do not provide sufficient protection.

One of the methods of protecting of titanium surfaces against corrosion is the preparation of non-metallic coatings, e.g. metal oxides. Due to the self-passivation of titanium, titanium dioxide is formed on its surface. The oxide formed on the surface of titanium provides corrosion protection and is the most common method of titanium protecting [8]. An oxide layer is formed on the titanium surface as a result of interaction with environmental factors. It can also be prepared by the anodizing method. This method allows for controlling parameters of the coating like e.g. thickness, which is impossible with self-passivation. Literature reports confirm the improvement of the corrosion properties of titanium after the preparation of a layer of titanium dioxide [9].

The addition of fluorine ions to the electrolyte used for anodizing allows to the preparation of titanium dioxide nanotubes (TNT) on the titanium surface. Titanium dioxide nanotubes provides better corrosion resistance of titanium than disordered titanium dioxide [10,11]. Titanium dioxide nanotubes are a porous coating with controlled TiO_2 thickness and nanotube diameter. This parameters improves the corrosion and protective properties of titanium [12].

The greatest advantage of choosing anodization as a method of titanium dioxide prepare is the ability to modify the process parameters so that the titanium dioxide is shaped like nanotubes. So far, the properties of single-walled titanium dioxide nanotubes have been extensively studied, but a novelty in recent literature reports is the possibility of producing double-walled titanium dioxide nanotubes [13–22]. Double-walled titanium dioxide nanotubes have an inner and outer wall, unlike TNT, which have a single wall [13,16]. Studies indicate a higher value of dTNT electrical resistance compared to TNT (lower conductivity), which may increase corrosion resistance of titanium [13]. Other dTNT properties are good photocatalytic activity and adsorption properties [14]. The good adsorption properties of the dTNT surface allow of deposition of additional components, which may increase the corrosion resistance of coating and give it a new properties. Unlike TNT, titanium dioxide double wall nanotubes retain their characteristic structure at temperatures above 600 °C. Therefore, it is possible to perform annealing of dTNT at temperatures higher than TNT, which are destroyed at too high a temperature [12,23].

The aim of the work was to improve the corrosion properties of titanium by developing a coating of double-walled titanium dioxide nanotubes on its surface. Only a few literature reports on the preparing and properties of dTNT can be found in the available sources. The first article on the method of dTNT preparing was published in 2009 by John et al. [17], but the properties and formation mechanism of dTNT are still little studied. The known properties of dTNT allow the conclude, that they can improve the titanium corrosion resistance.

2 Materials and Methods

2.1 Materials

Titanium (Ti) foil (Sigma Aldrich, purity of 99.7%, thickness of 0.25 mm), ethylene glycol (assay 99.8%), ammonium fluoride NH_4F, orthophosphoric acid H_3PO_4 (assay 85%) and sodium chloride NaCl. All solutions were prepared from distilled water.

2.2 Double-Walled TiO_2 Nanotubes Fabrication

The titanium foil was cut into 5×15 mm and sonicated in acetone and distilled water for 15 min, afterwards drying in nitrogen. dTNT layers were prepared by electrochemical anodization of titanium foils in 85% ethylene glycol solution with 0.65% wt. NH_4F and 7.5% H_3PO_4 in time 4 h using various potentials (10, 30 and 60 V) and a standard two-electrode system. Anodization was carried out at 0.05 V/s of speed of rise and using Autolab PGSTAT302N (Metrohm). A platinum sheet was used as a counter electrode and titanium foli as a working electrode (anodization surface 5×5 mm).

Scanning electron microscopy (FESEM, JEOL JSM-7600F) were used to investigate the surface morphology.

2.3 Electrochemical Measurements

The assessment of the corrosive properties was performed by: electrochemical impedance spectroscopy (EIS) scans, cyclic voltammetry (CV) measurements, open circuit potential (OCP) tests and potentiodynamic polarization. Ti and dTNT samples were recorded using a standard three-electrode configuration with a pure titanium sample and double-walled titanium dioxide nanotubes as the working electrode, the platinum foil as the auxiliary electrode, and standard silver chloride electrode ($E_{Ag/AgCl} = 0.222$ V) as the reference electrode.

EIS spectra were carried out over a frequency range of 0.1–10^5 Hz with a signal amplitude of 10 mV. Cyclic voltammetry and potentiodynamic polarization were performed for potential ranging from –1 to 1 V vs. Ag/AgCl with the scan rate of 0.05 V/s. OCP measurements were carried out for 14400 s at room temperature ($22 \pm 2\,^\circ$C). All measurements were made using 20 ml 0.9% NaCl as the electrolyte. All measurements were repeated five times.

3 Results and Discussion

Figure 1 shows SEM images of Ti A) dTNT with diameter of: B) 80 ± 10 nm, C) 200 ± 10 nm and D) 400 ± 10 nm prepared by using potential respectively: B) 10 V, C) 30 V and D) 60 V. It was observed the smooth surface of Ti (Fig. 1A). For the used anodizing parameters, the formation of dTNT layers on the titanium surface was found. Even though the dTNT structure isn't homogeneous, they are described as the double-walled titanium dioxide nanotubes according to previous literature reports [13, 15, 21]. It was observed that the higher the anodizing potential, the larger diameter of dTNT. The changes of dTNT diameter depending on the anodization potential are also confirmed by Motola et al. [13] for changes of potential in terms of 60–100 V.

The phase angle θ and impedance modulus —Z— recorded in the lowest frequencies (0.1 Hz) presented in Bode plots in Fig. 2A and B have decreased for Ti surface after prepared dTNT layer through anodizing. As the titanium anodizing potential decreases, the impedance values decreases, what was confirmed by Liu et al. [24]. The observed change in the value of the phase angle shows the improvement of the protective properties of titanium after anodizing [25]. In Fig. 2B two time constants were observed, which indicates the homogeneity of the surface of the tested sample and confirms the formation of an oxide layer. A shift in the frequency of the occurrence of time constants when using different potentials of dTNT formation was also observed, which may be related to the increase in the pores of the oxide, i.e. the diameter of the dTNT.

The Nyquist diagrams determined for Ti and dTNT layers shows fragments of wide, incomplete semicircles (Fig. 2C). The dTNT coating caused increase impedance ReZ of titanium. It also observed that increasing the anodizing potential increases the impedance value. The higher value of impedance means better stability and corrosion protection of surface, especially for dTNT prepared by potential 30 and 60 V [26]. Research by Motola et al. [13] have shown that dTNT has a higher impedance than single-walled titanium dioxide nanotubes and makes them potentially better corrosion protection for titanium.

Cyclic voltammograms recorded for Ti and dTNT layers are flat and shown any oxidation–reduction peaks in the potential ranging from–1 V to 1 V versus Ag/AgCl (Fig. 2D). dTNT layers shown higher separation between cathodic and anodic halwaves than titanium, in particular for potential −0.7 V. As the anodizing potential decreases, the separation between cathodic and anodic halwaves increases. This conclusion is the same for single-walled titanium dioxide nanotubes, what was confirmed by Arkusz et al. [12].

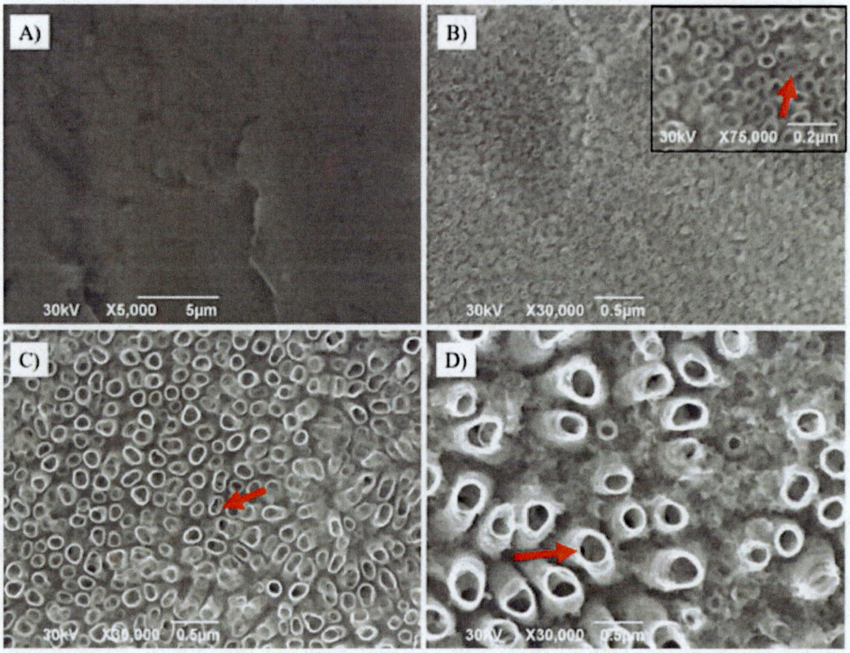

Fig. 1 SEM images of Ti **A** and dTNT showing the formation of dTNT using potential: **B** 10 V, **C** 30 V and **D** 60 V in 85% ethylene glycol electrolyte containing 0.65% wt. NH_4F and 7.5% H_3PO_4 in time 4 h. Examples of double walls of nanotubes are marked with arrows

Figure 3A presents the results of OCP values for Ti and dTNT layers. All measured samples shown stabilization of the OCP potential in time and no violation of passive layer. dTNT layer caused a decrease of corrosion potential of titanium. For a measurement time of 14400 s there was a dependence: the corrosion potential increases with the increase of the anodizing potential.

The corrosion current was determined by Tefel extrapolation from the polarization curve (Fig. 3B). For dTNT samples lack of symmetry of the anode and cathode branches of potentiodynamics polarization plots was observed. The results of the calculations of polarization resistance, corrosion potential and current were presented in Table 1. After prepared dTNT on titanium surface the corrosion current increased. The dTNT coating prepared at potential of 60 V also allowed to increase the polarization resistance. Increase of corrosion current and polarization resistance means better corrosion properties. Literature data confirm that the anodizing voltage increased improves the corrosive properties of the oxide layer in the presence of chlorine ions [24].

Table 1 Results of Tafel extrapolation of potentiodynamics polarization plots

	j_{cor} [A]	E_{cor} [V]	R_p [Ω]
Ti	$0.40 \cdot 10^{-6}$	−0.423	626677
dTNT10	$23.67 \cdot 10^{-6}$	−0.917	181303
dTNT30	$30.84 \cdot 10^{-6}$	−0.922	111450
dTNT60	$15.77 \cdot 10^{-6}$	−0.888	695371

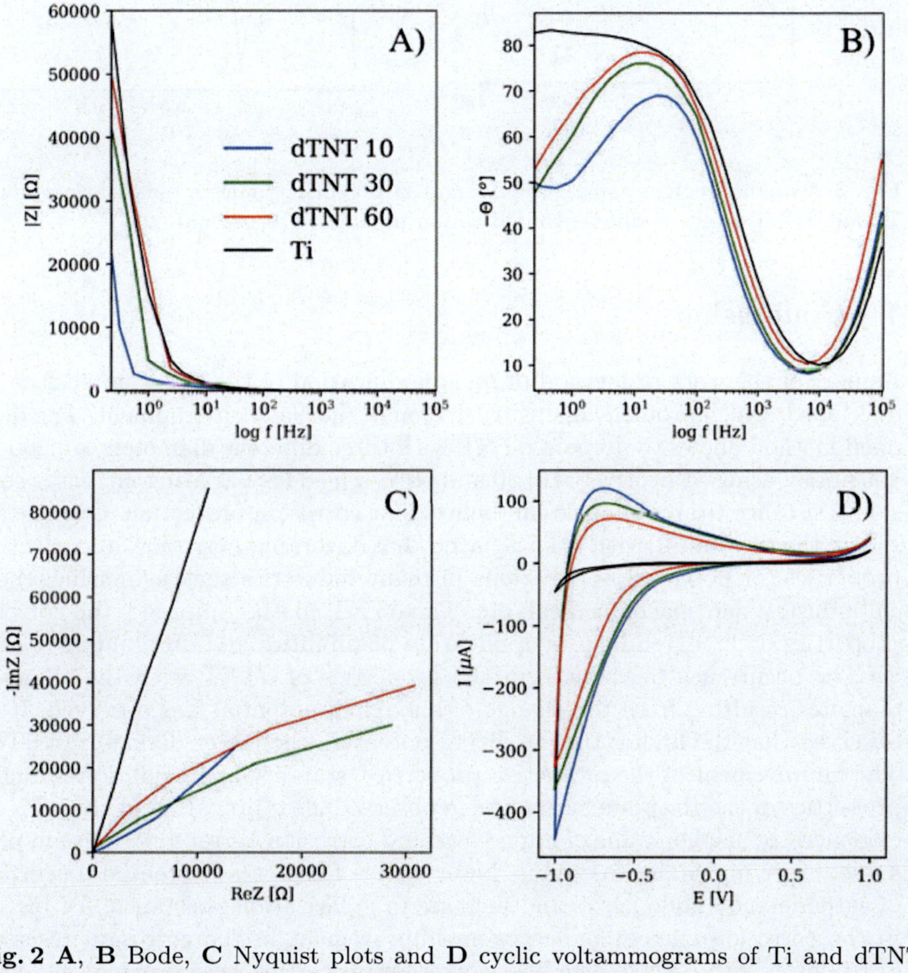

Fig. 2 **A**, **B** Bode, **C** Nyquist plots and **D** cyclic voltammograms of Ti and dTNT layers of different anodization potential (10, 30 and 60 V)

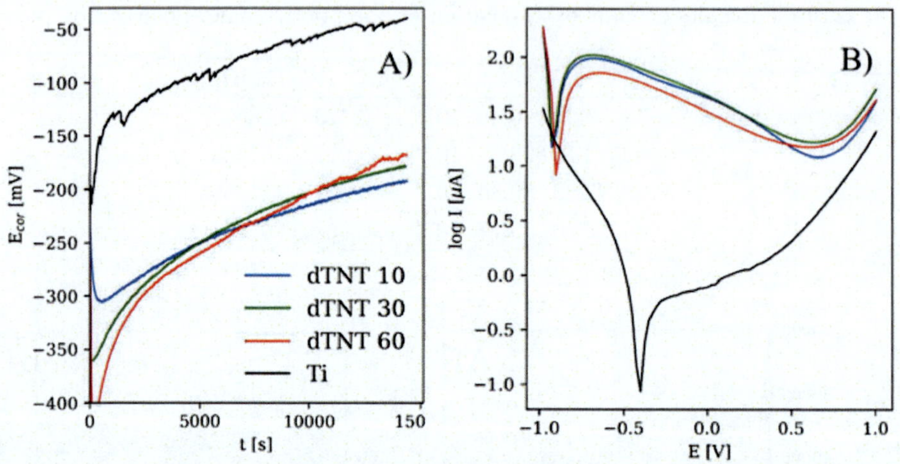

Fig. 3 **A** open circuit potential (OCP) and **B** potentiodynamics polarization plots of Ti and dTNT layers of different anodization potential (10, 30 and 60 V)

4 Conclusions

As part of the work, a method of nanomodification of the titanium surface was developed, which protects against corrosion in the NaCl environment. The developed method allows to prepare dTNT with three different diameters 80, 200 and 400 nm by using respectively 10, 30 and 60 V. The dTNT coating increases corrosion resistance titanium, so it can be used for corrosion protection. Our research shows that titanium with dTNT layers has favorable corrosion and electrical properties for potential applications in many industries such as marine, chemical industry, aerospace, medical, etc. The dTNT coating improves the corrosion properties of Ti, extending its application possibilities and durability.

The occurrence of changes in the properties of dTNT with the change in diameter resulting from the change of anodizing potential was observed. It was observed that the higher the anodizing potential, the larger diameter of dTNT. The improvement of the protective properties, stability and oxidation-reduction properties of the titanium surface were observed after preparing of the dTNT as evidenced by: higher value of impedance and corrosion current, decrease in phase angle, lower impedance modulus, lower value of corrosion potential, occurrence of cathode and anode peaks and increase in polarization resistance. An increase in the corrosion current indicates an improvement in the corrosion resistance of titanium with dTNT. The best corrosion properties were provided by dTNT layers with a diameter of 400 nm. Corrosion current for 400 nm dTNT increased by 15.37 μA, corrosion potential was decreased almost doubled, and polarization resistance increased by 68694 Ω compared to pure titanium. Potentially, the dTNT layer with a diameter of 400 nm is the best coating to prevent corrosion of titanium surfaces.

5 Summary

In this work, a nanomodification method of titanium surface was developed to protection against electrochemical corrosion. The aim of this study was to develop a new titanium coating that provides better protection against corrosion than an uncoated surface, as well as the characteristics of the formed coating. Anodization leading to the formation of double-walled titanium dioxide nanotubes was chosen as the nanomodification method. The developed method allows to obtain dTNT with three different diameters. The developed coating increases the corrosion resistance of titanium, so it can be used for corrosion protection. The as-prepared coating also has a number of other properties that can be used not only to improve the anti-corrosion properties of titanium, but also to impart new beneficial properties to the titanium surface.

References

1. Arkusz, K., Paradowska, E., Nycz, M., Mazurek-Popczyk, J., Baldy-Chudzik, K.: Evaluation of the antibacterial activity of Ag- and Au-nanoparticles loaded TiO_2 nanotubes. J. Biomed. Nanotechnol. **16**(9), 16–25 (2020). https://doi.org/10.1166/jbn.2020.29799
2. Arkusz, K., Paradowska, E.: Impedimetric detection of femtomolar levels of interleukin 6, interleukin 8, and tumor necrosis factor alpha based on thermally modified nanotubular titanium dioxide arrays. Nanomaterials **10**(12), 1–18 (2020). https://doi.org/10.3390/nano10122399
3. Williams, J., Boyer, R.: Opportunities and issues in the application of titanium alloys for aerospace components. Metals **10**, 705 (2020). https://doi.org/10.3390/met10060705
4. Goffredo, G.B., Accoroni, S., Totti, C., Romagnoli, T., Valentini, L., Munafo, P.: Titanium dioxide based nanotreatments to inhibit microalgal fouling on building stone surfaces. Build. Environ. **112**, 209–222 (2017). https://doi.org/10.1016/j.buildenv.2016.11.034
5. Meng, F., Yan, J.-Y., Olson, G.B.: TRIP titanium alloy design. In: MATEC Web Conferences, vol. 321 (2020). https://doi.org/10.1051/matecconf/202032111070
6. Dobrzański, L.A.: Podstawy nauki o materiałach i metaloznawstwo. 2. Wydawnictwa Naukowo–Techniczne (2002). ISBN: 8320427932
7. Myers, J.R., Bomberger, H.B., Froes, F.H.: Corrosion behavior and use of titanium and its alloys. JOM **36**, 50–60 (2013). https://doi.org/10.1007/BF03338589
8. Thomashov, N., Chernova, G., Ruscol, Y., Ayuyan, G.: The passivation of alloys on titanium bases. Electrochim. Acta **19**, 159–172 (1974). https://doi.org/10.1016/0013-4686(74)85012-7
9. Arkusz, K., Nycz, M., Paradowska, E.: Electrochemical evaluation of the compact and nanotubular oxide layer destruction under Ex Vivo Ti6Al4V ELI transpedicular screw implantation. Materials **13**(1), 1–16 (2020). https://doi.org/10.3390/ma13010176
10. Yu, W.-Q., Qiu, J., Xu, L., Guan, C.: Corrosion behaviors of TiO_2 nanotube layers on titanium in Hank's solution. Biomed. Mater. **4**, 12–65 (2009). https://doi.org/10.1088/1748-6041/4/6/065012

11. Arkusz, K., Krasicka-Cydzik, E.: The effect of phosphates and fluorides, included in TiO_2 nanotube layers on the performance of hydrogen peroxide detection. Arch. Metall. Mater. **63**(2), 765–772 (2018). https://doi.org/10.24425/122402
12. Arkusz, K., Paradowska, E., Nycz, M., Krasicka-Cydzik, E.: Influence of thermal modification and morphology of TiO_2 nanotubes on their electrochemical properties for biosensors applications. J. Nanosci. Nanotechnol. **18**, 13–21 (2018). https://doi.org/10.1166/jnn.2018.14685
13. Motola, M., Hromadko, L., Prikryl, J., Sopha, H., Krbal, M., Macak, J.: Intrinsic properties of high-aspect ratio single- and double-wall anodic TiO_2 nanotube layers annealed at different temperatures. Electrochim. Acta **352–479** (2020). https://doi.org/10.1016/j.electacta.2020.136479
14. Wang, Y., Zhang, X., Youa, S., Hu, Y.: One-step electrosynthesis of visible light responsive double-walled alloy titanium dioxide nanotube arrays for use in photocatalytic degradation of dibutyl phthalate. RSC Adv. **10**, 38–47 (2020). https://doi.org/10.1039/d0ra03627g
15. Chen, W., Yeh, M., Lin, L., Vittal, R., Ho, H.: Double-wall TiO_2 nanotubes for dye-sensitized solar cells: a study of growth mechanism. ACS Sustain. Chem. Eng. **6**, 07–15 (2018). https://doi.org/10.1021/acssuschemeng.7b04250
16. So, S., Riboni, F., Hwang, I., Paul, D., Hammond, J., Tomanec, O., Zboril, R., Sadoway, D., Schmuki, P.: The double-walled nature of TiO_2 nanotubes and formation of tube-in-tube structures and a characterization of different tube morphologies. Electrochim. Acta **231**, 721–731 (2017). https://doi.org/10.1016/j.electacta.2017.02.094
17. John, S.E., Mohapatra, S.K., Misra, M.: Double-wall anodic titania nanotube arrays for water photooxidation. Langmuir **15**, 40–47 (2009). https://doi.org/10.1021/la900426j
18. Ji, Y., Zhang, M., Cui, J., Lin, K., Zheng, H., Zhu, J., Samia, A.: Highly-ordered TiO_2 nanotube arrays with double-walled and bamboo-type structures in dye-sensitized solar cells. Nano Energy **1**, 796–804 (2012). https://doi.org/10.1016/j.nanoen.2012.08.006
19. Ji, Y., Lin, K., Zheng, H., Zhu, J., Samia, A.: Fabrication of double-walled TiO_2 nanotubes with bamboo morphology via one-step alternating voltage anodization. Electrochem. Commun. **13**, 1013–1015 (2011). https://doi.org/10.1016/j.elecom.2011.06.030
20. Ni, J., Noh, K., Frandsen, C.J., Kong, S.D., He, G., Tang, T., Jin, S.: Preparation of near micrometer-sized TiO_2 nanotube arrays by high voltage anodization. Mater. Sci. Eng. **33**, 259–264 (2013). https://doi.org/10.1016/j.msec.2012.08.038
21. Chen, S., Chen, Y., Li, Ch., Ouyang, H., Qin, S., Song, H.: Double-walled structure of anodic TiO_2 nanotubes in H_3PO_4/NH_4F mixed electrolyte. Mater. Res. Express **5**, 39–45 (2018). https://doi.org/10.1088/2053-1591/aabdd2
22. Hyeonseok, Y., Moonsu, K., Yong-Tae, K., Kiyoung, L., Jinsub, Ch.: Catalyst-doped anodic TiO_2 nanotubes: binder-free electrodes for (photo)electrochemical reactions. Catalysts **8**, 555 (2018). https://doi.org/10.3390/catal8110555
23. Albu, S., Ghicov, A., Aldabergenova, S., Drechsel, P., LeClere, D., Thompson, G., Macak, J., Schmuki, P.: Formation of double-walled TiO_2 nanotubes and robust anatase membranes. Adv. Mater. **20**, 4135–4139 (2008). https://doi.org/10.1002/adma.200801189
24. Liu, Z., Liu, X., Donatus, U., Thompson, G., Skeldon, P.: Corrosion behaviour of the anodic oxide film on commercially pure titanium in NaCl environment. Int. J. Electrochem. Sci. **9**, 58–73 (2014)

25. **Kierzkowska, A.**: **Wpływ** gięcia na charakterystykę in vitro anodowanej warstwy wierzchniej implantowego stopu tytanu Ti6Al4V ELI. Rozprawa doktorska. Uniwersytet Zielonogórski. Wydział Mechaniczny, Instytut Budowy i Eksploatacji Maszyn (2007)
26. **Sun, Z., Annergren, I., Pan, D., Mai, T.A.: Effect of laser surface remelting on** the corrosion behavior of commercially pure titanium sheet. Mater. Sci. Eng. **345**, 293–300 (2003). https://doi.org/10.1016/S0921-5093(02)00477-X